Women in Engineering and Science

Series Editor

Jill S. Tietjen
Greenwood Village, Colorado, USA

More information about this series at http://www.springer.com/series/15424

Pamela M. Norris • Lisa E. Friedersdorf
Editors

Women in Nanotechnology

Contributions from the Atomic Level and Up

 Springer

Editors
Pamela M. Norris
Department of Mechanical
and Aerospace Engineering
University of Virginia
Charlottesville, VA, USA

Lisa E. Friedersdorf
Department of Materials Science
and Engineering
University of Virginia
Charlottesville, VA, USA

ISSN 2509-6427 ISSN 2509-6435 (electronic)
Women in Engineering and Science
ISBN 978-3-030-19953-1 ISBN 978-3-030-19951-7 (eBook)
https://doi.org/10.1007/978-3-030-19951-7

This Springer imprint is published by the registered company Springer Nature Switzerland AG.
The registered company address is: Gewerbestrasse 11, 6330 Cham, Switzerland

Just as we were getting started with this project, we were deeply saddened by the passing of Dr. Mildred "Millie" Dresselhaus, perhaps one of the most impactful women scientists of all time. Beyond her technical contributions to science, she was an inspiration with her perseverance to overcome obstacles and with the passion with which she mentored students and faculty alike, including some of the authors included in this book.

Preface

Dr. Mildred "Millie" Dresselhaus was perhaps one of the most impactful women scientists of all time. In the spirit of this edition celebrating the contributions of women scientists and engineers to nanotechnology and telling their stories, a few brief details of Millie's inspiring life story are provided here. Authors also share memories of their interactions with Millie and the profound impact these interactions had on their careers.

Pamela Norris, coeditor, shares her first opportunity for significant interactions with Millie during the Seventh Japan-US Symposium on Nanoscale Transport Phenomena in Shima, Japan, in 2011: "We both arrived early, following the very long trip to Japan, and we were tired. But there was a dragon festival occurring that evening, and we were not going to miss the opportunity for first-hand observation. That evening as we soaked in the sights and sounds of Japan, I heard the amazing story of Millie's academic journey. I heard of her trials and hardships along the way, but I never heard a hint of resentment from a woman who had clearly blazed a new trial for women scientists. We discussed the challenges of balancing motherhood and academia, and I came to understand that the struggles along the path are just part of the journey. The science which motivates us, the contributions our science makes to improving the world and the human condition, and the privilege we have of working in a free world where the choices are ours to make and the path is ours to define, these are all things worth great work and effort and we are indeed blessed."

Millie was born in 1930 to immigrant parents and grew up in the Bronx during the Great Depression. A talented violin player, she attended a music school on scholarship. It was there she became aware of better educational opportunities and applied to Hunter College High School for Girls. She then attended Hunter College where she majored in liberal arts, but took as many science and math classes as she could. At Hunter, her physics instructor, Roslyn Yalow (who later became the first American woman to win a Nobel Prize for her work in the development of the radio-immunoassay technique), encouraged Millie to become a scientist. Millie saw a notice about a fellowship on a bulletin board, and after graduating in 1951, she was awarded a Fulbright Fellowship to spend a year working in the Cavendish Laboratory at the University of Cambridge in England. (The Cavendish Laboratory was established in 1874, first led by James Clerk Maxwell, to focus on experimental physics.) Millie earned her master's from Radcliffe College in 1953 and her PhD from the University of Chicago in 1958 where her dissertation focused on superconductivity. After a postdoctoral position at Cornell University, she joined Lincoln Laboratory as a staff scientist.

Millie was at Lincoln Lab from 1960 through 1967, a time when there was much focus on semiconductors. She focused, however, on semimetals, and her research led to early knowledge of the electronic structure of these materials, including graphite. In the 7 years she was at Lincoln Lab, Millie and her husband, Gene, had four children. The difficulty balancing four small children with a new laboratory policy mandating an 8 am start to the day led Millie to explore other opportunities. Millie joined the Department of Electrical Engineering at MIT on a fellowship from the Rockefeller Family Endowment aimed at promoting scholarship of women in science and engineering. One of the many notable contributions Millie made at MIT was changing the admissions criteria. The school had different standards for admitting women as opposed to men, "so the first thing that we did was to have equal admission that would be on the same criteria, academic criteria. That fact increased the women by about a factor two, just overnight." Millie spent the remainder of her career as a faculty member at MIT.

Millie's contributions to the advancement of scientific understanding are vast. Beginning with her work characterizing the electron band structure of carbon at Lincoln Labs, Millie became well known for her research studying the properties of carbon-based materials. She studied the effects of intercalation on graphite's electrical properties, in some cases inducing superconductivity. She also studied the novel properties of carbonaceous nanomaterials such as fullerenes and carbon nanotubes. Her experiments and calculations showed that the electrical properties were dependent on the chirality of the tube and that they can behave as a metal or a semiconductor (Dresselhaus et al. 1995). Due to her contributions to the field, including authoring four books on carbon, she earned the nickname, "Queen of Carbon." She is also credited with opening the field of low-dimensional thermoelectricity, as discussed in the chapter by Zebarjadi.

Millie was contributing author Mona Zebarjadi's co-advisor during her 3 years of postdoctoral studies at MIT. She has shared with us one of the many lessons that she learned from Millie. "The power of collaboration and net-working: Millie was constantly trying to put people in contact with each other. She was trying to bring scientists from different fields to sit in the same room and to share their ideas. She had a great vision over many fields as she served as an advisor for several government departments. She wanted her colleagues to reach out and expand their view points. In particular, she knew female scientists are not as well connected as male scientists, perhaps because of their much lower percentage. When I got my first job, she told me to establish my network, to go around and introduce myself to other faculties and make sure that I understand what they are working on even if I think their work is not relevant to mine. Later on, she followed up with me and asked me if I have found collaborators. Upon naming them, she said: 'Well, you are in good hands!'."

Although coeditor Lisa Friedersdorf did not have the opportunity to work with Millie, she reflects on her impact. "I recall attending presentations given by Millie at major conferences, the large room was always packed. I remember the ease in which she presented and answered, sometimes quite aggressive, questions. She was confident and the audience was differential as only in the presence of a well-respected scientist. Although not given to me directly, I often got second hand advice from those she mentored, men and women alike. 'Have you written your results?' They told me she'd say, always pushing them to do their best. I also heard a lot about collaboration, and helping others. I had the fortune to talk with Millie at the social time before her Kavli Prize Laureate Lecture at the Carnegie Institution for Science in 2013 and what struck me most was the personal attention she gave me, asking questions, listening, and offering encouragement. She was not only a world class scientist, but a genuinely nice person."

Contributing author, Evelyn Wang, fondly remembers Millie visiting her home while she was in middle school during Millie's visits with her father at CalTech. Millie always brought her violin with her, and Evelyn and Millie would play together. Evelyn shares "I hope that as we honor Millie, we remember both her scientific accomplishments—and her incredible way of putting those around her at ease. It's something I, personally, will never forget."

Indeed, in addition to advancing science, Millie was a passionate mentor and a well-known voice and an advocate for women in science. Her passion impacted generations of students at MIT and beyond. She was the recipient of many awards celebrating both her scientific and mentoring accomplishments, including the Medal of Freedom (the highest civilian award in the USA) bestowed by President Obama in 2014 and the Medal of Science awarded by President Bush in 1990. More information about Millie's exceptional life can be found at the sites listed below:

- American Physical Society. (2019). *Mildred "Millie" Dresselhaus*. Retrieved January 26, 2019, from http://www.physicscentral.com/explore/people/dresselhaus.cfm.
- Clean Energy Education and Empowerment Awards. (2012). *"Mildred Dresselhaus" Lifetime Achievement Award*. Retrieved January 26, 2019, from https://www.c3eawards.org/mildred-dresselhaus/.
- Dresselhaus, M. S., Dresselhaus, G., & Saito. (1995). Physics of carbon nanotubes. *Carbon, 33*(7), 883–891.
- Engineering and Technology History Wiki. (2013). *Oral-history: Mildred Dresselhaus*. Retrieved January 26, 2019, from https://ethw.org/Oral-History:Mildred_Dresselhaus.
- The Franklin Institute. (2017). *Mildred S. Dresselhaus*. Retrieved January 26, 2019, from https://www.fi.edu/laureates/mildred-s-dresselhaus.
- MIT Department of Physics. (2017). *Faculty, Mildred S. Dresselhaus*. Retrieved January 26, 2019, from http://web.mit.edu/physics/people/inremembrance/dresselhaus_mildred.html.
- MIT Infinite History. (2007). *Mildred S. Dresselhaus*. Retrieved January 26, 2019, from https://infinitehistory.mit.edu/video/mildred-s-dresselhaus.

- MIT News. (2017). *Institute Professor Emerita Mildred Dresselhaus, a pioneer in the electronic properties of materials, dies at 86*. Retrieved January 26, 2019, from http://news.mit.edu/2017/institute-professor-emerita-mildred-dresselhaus-dies-86-0221.
- U.S. News. (2012). *Queen of carbon science*. Retrieved January 26, 2019, from https://www.usnews.com/science/articles/2012/07/27/queen-of-carbon-science.
- Wikipedia. (2019). *Mildred Dresselhaus*. Retrieved January 26, 2019, from https://en.wikipedia.org/wiki/Mildred_Dresselhaus.

Charlottesville, VA, USA   Pamela M. Norris
                                                                          Lisa E. Friedersdorf

Note from the Editors

As part of the Women in Engineering and Science series, this book celebrates just a few examples of the many women who have advanced the field of nanotechnology. In the introduction, Friedersdorf and Spadola give a brief overview and use examples of contributions of women in a variety of technical areas impacted by nanotechnology. LeBlanc discusses how nanotechnology can be used to improve science and scientific literacy, differentiated by whether directed at a lay or technical audience. Some of the strategies used for incorporating nanotechnology in K–12 education are presented by Schmidt, with specific demonstrations and lessons for teachers and students.

The next few chapters take a more technical look at various aspects and applications of nanotechnology. Sayes makes a case for considering the environmental, health, and safety aspects of nanomaterials and gives the current state of understanding of nanotoxicology. Wen, Lee, and Steinmetz show how plant virus-based nanotechnologies are being developed for use in medicine, including for molecular imaging and drug delivery.

The impact of the nanoscale on thermoelectric transport properties of materials for applications such as waste heat recovery and solar power generation is discussed by Zebarjadi. Wang, Zhu, Mutha, and Zhao consider how nanostructured surfaces can control the manipulation of water for thermal management, energy production, and desalination. Norris and Larkin review developments in the understanding of thermal boundary resistance with a view toward the future when the ability to engineer the interface in order to optimize thermal performance will be possible.

Finally, Merzbacher provides a historical perspective on the US National Nanotechnology Initiative and provides her perspective on future prospects for nanotechnology.

Contents

About the Authors

Lisa E. Friedersdorf A piece of advice I often give my students is to build a strong foundation so they can take advantage of opportunities as they arise, and I have been fortunate to be able to do so myself. I never could have foreseen the professional path I've taken nor charted a course to the position I now hold.

I graduated summa cum laude with a bachelor's degree in mechanical engineering from the University of Central Florida. While there, I had the opportunity to conduct independent materials science research, an experience which was pivotal to everything that followed. Not only did I learn laboratory skills, but I also assisted with the finances and writing reports and proposals. The research group of graduate and undergraduate students was dynamic and collaborative. Some of my favorite memories are the all-nighters making posters in the days leading up to major conferences. My research focused on electrochemistry, to study the environmental degradation of high-temperature superconductors and stress corrosion cracking of stainless steel alloys. For graduate school, I wanted to study surface oxidation with scanning probe techniques, tools that were just becoming commercially available. I wrote a proposal to NSF on this topic and was awarded a 3-year graduate fellowship. I joined a group at the Johns Hopkins University to pursue a PhD in materials science and engineering. That fall, I married a doctoral student also in materials science. The delivery of the scanning probe system was delayed so I worked on a side project to fabricate and characterize photoluminescent porous silicon which became the topic of my master's thesis. As I was finishing the required coursework and ramping up my doctoral research, several unexpected events took place. I had a difficult pregnancy that required months of bed rest, my advisor left the university, and my husband took a job on the other side of the country at the US Bureau of Mines in Albany, Oregon. I knew I wanted to teach at the university level which required a PhD, so after relocating with a newborn, I worked to find a way to finish my degree. Having completed my coursework, the first few months were focused on studying for the graduate board exam which gave me time to look for the resources to do my research. The Bureau of Mines let me use a small concrete building where I could control vibration, Hopkins let me borrow the scanning probe system, Linfield College gave me access to equipment in the physics department to make samples, and I used the library at Oregon State University. The lessons learned in building partnerships were perhaps even more valuable than the advancements I made in understanding of the initial stages of copper oxidation using scanning tunneling and atomic force microscopy. Meanwhile, congressional initiatives called for the elimination of the Bureau of Mines, so my husband accepted a position at Bethlehem Steel's Homer Research Labs in Pennsylvania. Since this was before the widespread use of email, my advisor and I spent hours on the phone reviewing drafts of my dissertation sent by FedEx. I remember clearly pulling the final page of my dissertation off the printer, dropping keys on my supervisor's desk, pulling the gate shut, and jumping in the car on our son's second birthday for the trip back East.

Once settled in Bethlehem, I took a part-time research position in the Materials Research Center at Lehigh University working on functionally graded thermal barrier coatings for turbine blades until I had my second son. Although mostly at home for the next year, I worked with Lehigh faculty to write proposals and papers, and then I got a call from the center director. He offered me an industrial liaison position with the promise of flexible hours; it was the perfect fit. I found I was well suited to promoting and managing the complex interactions between academia and industry. Over the next 5 years, I grew the program, worked closely with Ben Franklin to support entrepreneurs and local start-ups, and established a multimillion dollar public–private partnership in microelectronic packaging. I also built my own research group, funded by the Office of Naval Research, and taught corrosion and electrochemistry classes for the materials science department. The steel industry, however, was not doing well and we decided to do a nationwide job search shortly before Bethlehem Steel declared bankruptcy.

Finding positions for two professionals is always a challenge, but especially so when you have the same technical specialty. I was offered two faculty positions and my husband had offers from a small company and a research institute. Although I loved teaching and research, at that time I decided that a tenure-track faculty position would not allow me the flexibility I still wanted with my small children, which was certainly a deviation from my original professional plan. We moved to Charlottesville, Virginia, where my husband joined a small company. For the first year, I continued to work for Lehigh managing the public–private partnership and helping to revise the university's intellectual property policy. I became actively engaged in the tech community and began consulting for the Virginia Center for Innovative Technology as my responsibilities at Lehigh wound down. I also took a position teaching physics and advanced placement chemistry at a local private high school. My role at CIT grew and I was promoted to director of the Virginia Nanotechnology Initiative where I led an alliance of academic institutions, industry, and government laboratories. I worked closely with the legislative and executive branches of the state government and the Virginia congressional delegation. I compiled an inventory of nanotechnology assets; facilitated research and commercialization collaboration; prepared and presented reports, strategic plans, competitive analyses, and investment proposals; reviewed legislation; provided technical support and briefings; and conducted outreach to build community. During this time, I was also building connections to the Office of the Vice President for Research at the University of Virginia and they created a new part-time research program manager position. In this role, I was responsible for building cross-school teams and leading proposal development for large programs. I also supported statewide efforts including the Virginia Research and Technology Advisory Commission (VRTAC) and served on Joint Commission on Technology and Science (JCOTS) citizen advisory committees. Although I was juggling family and several part-time jobs, I had the flexibility to work around the clock. As funding for the VNI was declining, I moved into a full-time position at UVA with an equal split between the program management role and as managing director of the Institute for Nanoscale and Quantum Scientific and Technological Advanced Research (nanoSTAR). At nanoSTAR, I oversaw institute operations including research program development, budget management, marketing, communications, and outreach. I established and administered funding programs for faculty seed projects, undergraduate summer research, and graduate student travel. I also designed and established an industry collaborative research program and facilitated the formation of a multimillion dollar public–private partnership in nanoelectronics. My passion for mentoring students continued and I taught materials science and corrosion classes, advised numerous senior projects, served on graduate committees, and was the advisor to Tau Beta Pi and the Nano and Emerging Technologies (NExT) club.

Shown an advertisement clipped from "The Hill" for a policy analyst, I reached out to colleagues at the National Nanotechnology Coordination Office (NNCO) to learn more. With one son in college and the other finishing up high school, I became a consultant at the NNCO while maintaining a visiting position in materials science at UVA. My primary responsibility was to support the Nanotechnology Signature Initiatives and after a couple of years I was asked to join the leadership as deputy director. I returned to UVA as a principal scientist and under the intergovernmental personnel act (IPA) joined the National Science Technology Council of the White House Office of Science and Technology Policy, and was later promoted to Director. In this role, I now lead the office that provides technical and administrative support for the National Nanotechnology Initiative and coordinates collaboration among the 20 federal agencies that invest approximately $1.5 billion annually in nanotechnology research and development. I also serve as the spokesperson for the NNI nationally and internationally. In addition to facilitating interagency coordination, I have strengthened communication with the research community through collaboration with technical societies, associations, and major conferences; initiated and expanded mechanisms for public outreach and STEM education including podcasts, contests, videos, animations, and student and teacher networks; and expanded the use of communities of interest including the US-EU communities of research.

My technical knowledge is of course an asset that enables me to advance major R&D initiatives, but my experience communicating across sectors and building collaborations has been equally important throughout my career. And it's a lot of fun.

 **LeighAnn S. Larkin** Retrospectively, my path into science was predictable. I was a curious child, and in school I had always loved my math and science courses. However, it was not until my senior of high school that I made the conscious decision to begin my journey to becoming a scientist. At 17 years old, with the encouragement of my high school physics teacher, I participated in a research opportunity during which I was able to conduct my own research. I immersed myself in the sciences and solidified my goal to eventually become a scientist. I started my journey majoring in physics at a small liberal arts school in NY state, The College at Brockport. In my senior year of college, I took my first quantum mechanics course. The strange and complex laws governing the world of the very small intrigued me more than any topic I had previously studied. I was fascinated by how these quantum mechanical properties influenced the macro-world we live in. I observed as modern technologies were utilizing nanoscale phenomena to greatly improve our quality of life. I decided I wanted to pursue a PhD where I could study fundamental principles to improve our world's nanotechnology. Upon graduation, I immediately joined the Nanoscale Heat Transfer laboratory under the advisement of Dr. Pamela Norris at the University of Virginia. I began studying how to best optimize heat transport for a range of applications, such as thermoelectric, magnetic storage, and microelectronic devices. I am currently in the process of finishing my PhD in Engineering Physics and finishing a dissertation aimed at advancing our current understanding of thermal transport across metal/semiconductor interfaces. As I progress in my career, I hope to continue conducting research on the transport properties of materials and how these properties can be tailored to improve the technology ubiquitous in our everyday lives.

As a first-generation college student, I believe that without the encouragement and support of my mentors and peers, my journey may have evolved very differently. I have channeled these beliefs into a commitment to mentoring the next generation of scientists and improving the climate of the sciences within the academy. I joined UVA's NSF-sponsored Institutional Transformation ADVANCE program as a research assistant. The ADVANCE program is focused on methods to increase representation of women in the academy in the sciences. The UVA program, referred to as UVA Charge, has a special emphasis on Voices and Visibility, increasing the visibility and sense of belonging of STEM women at UVA. My research has been focused on collecting ethnographic data from UVA's staff and faculty on the diversity climate at UVA and how they have been influenced by institutional change programs and policies. In addition to following my own career path, I believe that it is equally important to create an environment that enables our next generation of scientists to be able to pursue their own passions and craft their own journey.

Saniya LeBlanc My engineering career path has been a windy road with detours down avenues in education and service. For most of that path, my identity has been one defined by exception. The first exception was my decision to be a mechanical engineer. Unlike most people, I chose my discipline when I was a young child, and I never changed my mind (or regretted the decision). My upbringing was a fortunate exception. From a family full of female STEM professionals, I never realized that my career choice might be atypical, my proficiency in math might defy a societal norm, or my gender might distinguish me. I passed most of my undergraduate career in happy oblivion, learning independently with my nose in a book, reveling in the beauty that is engineering, vaguely aware that about 80% of my peers were male and only one of my engineering classes was taught by a woman (a fact which held true for the duration of all four of my engineering degrees).

Some very determined professors nudged me towards graduate school and graduate fellowships, so, with a B.S. in mechanical engineering (and a minor in French) from the Georgia Institute of Technology, I went to the University of Cambridge as a Churchill Scholar to earn a research-based master's degree in engineering. I had deferred admission to Stanford University, and I was supposed to head to sunny Palo Alto to pursue a PhD after finishing my degree at Cambridge. My path seemed clear, but my heart was going in another direction. I felt a calling to serve society in a meaningful way, and it was not clear how graduate school fulfilled that calling. Maybe it was the many hours in a cleanroom bunny suit trying to fabricate my device to no avail, but research was not fulfilling my passion for engineering.

Skimming a bookshelf in a college bar (the college bars have libraries at Cambridge), I picked up *Savage Inequalities: Children in America's Schools* by Jonathan Kozol, and, for the first time, I stared my privilege in the face. The systemic inequity in our education system appalled me. I was humbled ... and impassioned to become an educator. I passed up Stanford's graduate fellowship, deferred the National Science Foundation Graduate Research Fellowship, and joined Teach For America to teach math and physics in an urban high school—to the dismay of family, friends, and mentors who were concerned it was a career mistake.

Teaching in a diverse, high-needs community fundamentally altered my understanding of society. With graduate classes in education, professional development, and many, many hours of exhausting practice, I started to learn how to be a teacher. More importantly, I started to understand how policy, paradigms, culture, and social injustice form a tangled web in which so many people get caught. I was no longer able to wrap myself in a cocoon of science and engineering, and my career path as an educator, engineer, or something else entirely was unclear. Although I was deeply fulfilled by my role as a teacher, I missed engineering. I longed for that hard engineering that makes you feel exhausted but energized when you finally conquer a technical challenge. I even missed the clean room (sort of). Maybe those mentors were right, and I should be an engineering professor—an engineer and an educator.

After a 3-year detour, I finally made it to Stanford University where I felt very much like an exception. I had never heard of Stanford before an undergraduate professor told me to apply there, so maybe it is no surprise I was out of place. Perhaps out of a sense of survival, I focused on what I knew best—learning. I relished the cornucopia of engineering topics—nanomaterials, energy systems, thermal transport, and microsystems—until I found the theme which united it all: energy. It links the nanoscale to the macroscale and provides invaluable services to society. The materials, manufacturing techniques, devices, and even economics of energy systems fascinate me. I also used the time at Stanford to learn about engineering education, especially the education research which drives better teaching and learning.

After obtaining my PhD, my passion for energy technologies led me to join Alphabet Energy, an energy technology startup company, as a research scientist. I created research, development, and manufacturing characterization solutions for thermoelectric technologies and evaluated the potential of new power generation materials. It was an outstanding industry experience, but the educator in me missed working with students.

In 2014, I joined the Department of Mechanical and Aerospace Engineering at the George Washington University. With a grant from the National Science Foundation, I created an undergraduate Nanotechnology Fellows Program which allowed me to combine my research and education expertise to influence future generations of scientists and engineers. The experience prompted the discussion presented in this chapter. My research is also an enjoyable exception since it spans many disciplines: thermal sciences, materials science, mechanical engineering, engineering economics, and engineering education to name a few. I have the privilege of conducting exciting, interdisciplinary research in advanced materials and manufacturing techniques for energy systems. I hope to push the boundaries of how we think about the link between materials, manufacturing, and systems for energy technologies.

As I continue on my path, I hope to use my experience as the exception to serve others and offer a perspective that deepens discussions between scientists, engineers, and educators. Many people do not have the luxury of learning independently, immune to societal pressures, identities, and stereotypes. They are not handed the privilege of unabated educational opportunity with the chance to pursue a career about which they are passionate. I aim to recruit and guide future engineers through the perilous journey of finding their own exceptional paths.

Karin L. Lee My interest in pursuing biomedical research started at a young age. As long as I can remember, I was interested in science; biology and chemistry were my favorite classes throughout grade school. Toward the end of high school I decided that I wanted to pursue a career in biomedical research after attending an engineering summer camp and participating in a summer internship in a biomedical engineering lab. I found the field appealing because the research done was directly able to impact human health. However, as an undergraduate, I learned that the field of biomedical engineering was broader than I had imagined and encompassed a range of research interests, everything from prosthetics and implants to tissue engineering and nanoparticles. Luckily, I had the opportunity to work in multiple labs that had different research interests and found that I was most interested in using biomaterials for biomedical applications.

My specific interest in nanotechnology grew out of this interest in biomaterials. I was first introduced to nanotechnology as an undergraduate, and as I considered graduate schools, I aimed to find a lab that worked with nanoparticles. I ultimately joined Dr. Nicole Steinmetz's lab, where we focused on using plant viruses as nanoparticles. I was fascinated by this work because the concept of using natural carriers to better human health was unique and innovative. My thesis work focused on the development of potato virus X (PVX), a flexible, elongated plant virus, for use as a cancer therapy. My early projects with PVX utilized it as a traditional nanoparticle for drug delivery, while my later projects investigated its use for immunotherapy. I received my PhD from Case Western Reserve University in 2016 and soon after started my postdoctoral research.

I am now a postdoctoral research fellow at the National Cancer Institute within the National Institutes of Health. I am working on the development of cancer vaccines, to be used in combination with other immunotherapies, and have had the opportunity to learn many new techniques. I am lucky to have been able to take my interests in nanotechnology and biomedical engineering and apply the skills I learned toward preclinical research in new in vivo models, which will be used to guide clinical trials. As I move forward in my career, my hope is that the work I've done and continue to do will contribute to improving clinical options for cancer patients.

Celia Merzbacher I have always been intrigued by understanding the fundamental physics and chemistry of materials. My doctoral research was in geochemistry, investigating the arrangement of atoms in silicate melts, aka magmas. By the time I graduated, my interest had shifted. I was attracted to work at government laboratories, where I engaged in basic research on man-made materials to address real-world problems.

I began my career as a postdoctoral fellow at Lawrence Livermore National Laboratory and then took a position as a researcher at the US Naval Research Laboratory (NRL) in Washington D.C. My research focused on developing and characterizing novel materials using various spectroscopic and microscopy techniques. I performed early research on high-temperature superconductors to better understand the structure of oxygen vacancies, which correlate with the temperature at which superconductivity is observed. However, most of the materials that I investigated were amorphous. I had the opportunity at NRL to work on materials for IR-transmitting optical fibers, space-based optics, thermoelectric devices, nonreflective coatings, and IR decoys. Working on materials that could make a difference—for the Navy and for society—was very rewarding.

In 2003–2008, due in part to my expertise in nanoscience, I was asked to serve at the Office of Science and Technology Policy (OSTP) in the White House where I oversaw the expanding activities under the National Nanotechnology Initiative (NNI). I co-chaired the interagency Nanoscale Science, Engineering and Technology (NSET) subcommittee and worked closely with the National Nanotechnology Coordination Office. Nanotechnology was a research priority while I was at OSTP and the multiagency investment grew to nearly $1.5 billion. Important steps taken in the years following the enactment of the Twenty-First Century Nanotechnology Research and Development Act in 2003 included development of the first NNI strategic plan to guide the program, establishment of an interagency Nanoscale Environment and Health Implications (NEHI) working group to identify and mitigate risks of the emerging materials and products, launching of nanotechnology standards activities, and broad engagement with industry and other stakeholders. I coordinated reviews by the President's Council of Advisors on Science and Technology (PCAST) published in 2005 and 2008. The framework established for the NNI has continued to support the program as it evolves and matures and has been adapted to the National Quantum Initiative launched in 2018.

Following my tenure at OSTP, in 2008 I became the Vice President for Innovative Partnerships at the Semiconductor Research Corporation (SRC). SRC is a consortium of the semiconductor industry that invests in basic, precompetitive research at universities to address the long-term needs of member companies, often in partnership with federal agencies. Today's microelectronics depend on nanostructured materials, making the semiconductor industry, in a sense, the largest nanotechnology-based industry. At SRC I worked with member companies to identify and develop areas of research that would enable the industry to continue to create new nanotechnologies and co-authored in 2016 a vision and research guide for "rebooting the IT revolution."

In 2019 I joined the Quantum Economic Development Consortium (QED-C), an industry consortium that was established by the National Quantum Initiative Act signed in 2018. QED-C is supported by the National Institute of Standards and Technology (NIST) with the goal of growing a robust US quantum industry. Quantum technologies are enabled in many ways by advances in nanotechnology and offer myriad applications in sensing, communications, and computing. The consortium, working in partnership with government, aims to ensure that the United States remains a leader in this critical and economically important area of technology.

My work in nanotechnology also led to opportunities to serve as advisor to various efforts and organizations. I was on the committee that performed the National Academies triennial review of the NNI in 2013 and I chaired the committee that performed the review in 2016. I have helped to review and advise nanotechnology programs and projects funded by the National Science Foundation (NSF) and the Department of Energy (DOE). I am on the advisory board of the SouthEast Nanotechnology Infrastructure Corridor (SENIC) co-led by the Joint School for Nanoscience and Nanoengineering and Georgia Institute of Technology. In addition, since 2009 I have served on—and chaired since 2012—the advisory board of the Penn State Nanotechnology Applications and Career Knowledge (NACK) center. Funded by NSF, the center has focused on educating students in 2-year programs, e.g., at community colleges. An important contribution of the NACK center has been the development of ASTM guidelines for 2-year curricula to ensure that students graduating from such programs are equipped with fundamental nanotechnology knowledge and skills to meet industry needs. It has been an honor to be part of the advancement of nanotechnology through my roles as advisor and reviewer.

My career in nanotechnology has spanned basic scientific research, policymaking at the White House level, and establishing and guiding innovative R&D and workforce development programs. I am passionate about crossing boundaries among disciplines, research institutions, sectors, government agencies, and even nations. Nanotechnology both benefits from and is a vehicle for the creation of such diverse connections. I hope to continue to break down silos and remove friction in the nanotechnology research and education systems, so that society can realize the benefits of our ability to control matter at the scale of atoms and molecules.

Heena K. Mutha In high school, at the Illinois Mathematics and Science Academy (IMSA), I was taught how to learn through scientific inquiry: pose questions, conduct experiments, and analyze results. I enjoyed research and loved testing hypotheses, and I wanted to use science to solve global challenges. I decided to pursue an undergraduate degree at Franklin W. Olin College of Engineering, where the design process is used to develop targeted, context-based solutions. At Olin as a mechanical engineering major, I was able to work on a variety of projects at the macro- and microscales: building an automated microscope platform for microfluidics research, designing a liquid crystal display by actuating crystals under an electric field, building a hovercraft with a lawnmower engine, and more. I was especially interested in thermodynamics and mass transport as it could be used for developing renewable energy technologies. While at Olin I interned at the National Renewable Energy Laboratory in Boulder, CO, under the DOE undergraduate research opportunity to do dynamic wind turbine modeling for extending lifetimes of turbines. These undergraduate projects and research experiences framed my desire to pursue a graduate degree in mechanical engineering.

In parallel, I also had spent both high school and college developing and running engineering outreach programs for K-12 students. This allowed me the opportunity to spend a year in India, studying the scalability of science informal education programs. While in India I observed that potable water was a scarce resource in some rural and urban communities, whether it was only provided by the municipal for a few hours a day, or transported by tanks and collected by people standing in line with empty pots to fill. There is a large gap worldwide for delivering potable water, where engineering solutions will be key to solving this challenge. I decided to use my dissertation research as an opportunity to focus on low-energy water desalination solutions.

At the Massachusetts Institute of Technology I conducted graduate research in water desalination using nanomaterials. High surface area, nanostructured electrode design, and device integration are key to scalable capacitive deionization systems. Nanoscale engineering and surface chemistry can alter desalination performance (throughput, water quality, lifetime) at the device scale. My work at MIT has given me the tools to continue to use nanoengineering to build solutions in water, renewable energy, and beyond. At present, I am a senior member of the technical staff at the Charles Stark Draper Laboratory in Cambridge, MA. My research includes next-generation electrochemical energy storage, biosensors, microfluidic platforms for high-throughput immunotherapy, and nanowire manufacturing. I hope to use my background in mechanical engineering and nanotechnology to continue to innovate solutions for water, energy, and biotechnology. In addition, I hope to continue to mentor and share that passion with young women and students to pursue a career in engineering.

Pamela M. Norris I was in fourth grade in Portsmouth, VA, when my class visited a lab sponsored by NASA meant to introduce kids to the field of engineering. At the conclusion of the exhibit, there was a "computer" that asked the question what two topics lead to a career in engineering. I guessed my two favorite topics, math and science, and was rewarded a little patch that said "engineering." I went home and told my mom "I want to be an engineer when I grow up." As a single mom and in a family where no one had attended college, she had no idea what an engineer was, but she instilled in me the confidence that I could be whatever I wanted. The best gift she could have given. I decided then I wanted to major in engineering with the ultimate goal of becoming a math and science teacher, probably in high school.

I stayed home and attended Old Dominion University for my undergraduate degree in mechanical engineering, as it was the only affordable option available to me. Last year, only 13.5% of degrees in mechanical engineering nationwide were awarded to women. I'll never forget the first day of my thermodynamics class. As I sat eager and bright-eyed on the front row next to my friend, Diana, the instructor looked out and said: "Ugh, girls are not supposed to be engineers." I'm pretty sure his words did not have the effect he intended. For I am certain he could not have said anything that would have motivated me more. At the end of the semester, there were two As assigned, to the only two females in the class, and the instructor would later write me one of my strongest letters of recommendation for graduate school. This is a powerful example of how people's words can affect us—but the effect is really a choice we make in how we interpret and process the words.

Contrast this to my fluids teacher. He made the mistake of writing, "Seems you are not doing quite as well in this class as everyone thought you would." His words also did not have the effect he intended. I was infuriated that others were discussing their expectations of me—in this case, their expectations that I would be one of the best students in the class. I responded by receiving one of the few Bs I got as an undergraduate. I really resented others expressing preconceived notions of my abilities—I didn't want to be prejudged, good or bad.

Nearing the end of my bachelor's degree, I was prepared to start looking for high school teaching positions when the department chair, Dr. Bob Ash, approached me to encourage me to consider graduate school. That thought really was outside of my comfort zone, and I don't think I ever would have gone to graduate school without his encouragement. So, I applied to MS programs (figuring with a MS, I could potentially teach courses at a community college). I applied to MIT, my dream school, and Georgia Tech, my backup. Dr. Bill Wepfer, the graduate director and my ultimate advisor, called me multiple times to convince me that Georgia Tech was the best choice for me. It was Dr. Wepfer who convinced me I was capable of a PhD. (This is when my career goals changed again, and I decided that a faculty position at a teaching college would be my goal.)

The PhD program had its own share of bumps along the way. My research was in the rather traditional and extremely male-dominated field of heat transfer in diesel engine cylinder heads. Perhaps the most significant bump along the way, however, was the PhD qualifying exam. The first try I aced thermodynamics but failed fluids and heat transfer. Then the second time I aced heat transfer but failed fluids. (Remember, I chose to get a B in fluids at ODU.) I was supposed to be released from the program, but Dr. Wepfer went to bat for me and I was allowed a third attempt. In the 6 months of prep I took five fluids classes, and yes, aced fluids the third time around. Those 6 months were definitely an emotional low for me, but it made me really examine what I wanted—YES, I wanted a PhD and to teach at the college level. And this story of failure has served me quite well as an advisor and mentor when students need to hear of experiences overcoming adversity and of failures.

As the end of my PhD approached, I began interviewing for faculty positions at mostly teaching colleges. I had several offers immediately, but from places that did not really excite me. Then I had the good fortune of meeting the then chancellor of Berkeley, Dr. Chang-Lin Tien—the father of the field of microscale heat transfer. He was giving a distinguished lecture on the topic of "excellence through diversity." He truly believed that only with a diverse team can you have creative solutions to engineering problems and that better solutions really result from a diverse set of people talking about the issues and working collaboratively.

At a reception following his talk I told him I was about to graduate as only the third woman ever to get her PhD in mechanical engineering at Georgia Tech and he said, "Oh, you should come do a post-doc with me." So after a few weeks I called to tell him I'd like to accept his offer and his response was, "Oh, I'm sorry, I don't really have a position, but good luck." While deflated, I thought more about it and I called him back and said, "I really want to come do a post-doc with you. Perhaps I could teach a class, or maybe find my own funding. I just want to work with you. I'll work for free." He was left speechless. He called back 2 days later and said he'd found a position for me. Definitely the best career move of my life because he was an amazing mentor. He approached each mentoring situation differently—recognizing that mentoring is not a one-size-fits-all activity.

I also totally switched my field of research while at Berkeley, from heat transfer in diesel engines to the much more exciting, cutting-edge field of microscale (or nanoscale) heat transfer. This bold move required a huge dose of self-confidence! Importantly this new field was very supportive and nurturing. Nearly all the faculty within the field at that time traced their foundation to Tien (most still do) and he had taught them a culture that was very supportive. In this community mentoring is expected and valued.

When I left Berkeley I ultimately chose UVA over a few other offers. I have been fortunate enough to have been given many unique opportunities along the way. The then provost Gene Block sponsored my attendance at a month-long intensive residential program for Women in University Administration in 2007—a clear signal of the University's commitment to my development. I was, however, quite aware that I was often the only female in the room, and I've often found myself in the position of having to be the spokesperson or the watch guard for "all women" in similar minority populations. Wishing to enhance the environment for other women in STEM, I began applying for an NSF Institutional Transformation grant in 2005, and on our third attempt we were finally awarded a $3.2 M ADVANCE IT grant in 2012. I'm extremely proud of the work we are just now finishing up with this award.

When I completed a Mission Writing workshop a few years ago, I converged to "My mission is to help others, be they my kids, students, faculty, or co-workers, to succeed beyond their wildest dreams." It is this mission which motivates. As I have accepted additional administrative responsibilities, first as Associate Dean of Research and Graduate Studies in 2011, then as Executive Associate Dean for Research in 2015, and as Executive Dean in 2018, it is because the ability to potentially impact a larger sphere of people excites me.

Christie M. Sayes People often ask me how and why I became a scientist. My path is not traditional, but it was clear to me; I lived by "seize the opportunity when it presents itself." I attended Louisiana State University in Baton Rouge, Louisiana, in my college years. While those years were certainly transformational in character, it was the next 4 years that shaped my future and placed me on the path where I am today. I attended Rice University in Houston, Texas, during my graduate school years. As a trainee in the chemistry department, I was challenged in physical chemistry (e.g., quantum mechanics, thermodynamics, wavefunctions, and heat transfer). But I enjoyed in the adversity; after all, nothing worth doing is easy. Not only could I learn the subject matter, but I could also communicate it (as a teaching assistant) and practice it (as a research assistant). I learned that being a trainee meant more than being a consumer of information; I also had the responsibility to produce new knowledge for the scientific community. I learned quickly that criticism was part of the job and accolades are few and far between. In my experience, both (critique and praise) have the potential to thrust you into the next realm of scientific curiosity—which corresponds directly with professional success.

I am now a practicing research scientist in the fields of chemistry and environmental health. Currently, I hold the position of Associate Professor of Environmental Science and Toxicology at Baylor University in Waco, Texas.

The 2018 Sayes Research Group in the Department of Environmental Science at Baylor University: Students studied issues related to nanotoxicology, nanomedicine, and particle chemistry. Top row (left to right): Gaby Cruz, Thelma Ameh, Henry Lujan, Daniel Kang, and London Steele. Bottom row (right to left): Sahar Pradhan, Desirae Carrasco, Christie Sayes, and Marina George

My subject matter expertise includes advanced materials, human exposure and health effect exposure, and risk science. My activities include working with partners, collaborators, and clients in designing and directing studies and training and advising facility staff. I possess a working knowledge of laboratory science and US regulatory climates. Routine activities include data collection, analyses, and interpretation as well as result documentation and reporting. Data is always related back to the published literature. Lastly, I function as point of contact for study control. Formerly, I served as a director of the environmental health program at RTI International and an Assistant Professor of Toxicology at Texas A&M University.

The 2010 Sayes research group in the Department of Physiology & Pharmacology at Texas A&M University: Students studied the mechanisms of toxic action after engineered nanomaterial exposure. From left to right: Michael Berg, Aishu Sooresh, Christie Sayes, Niveeta Bangeree, and Amelia Romoser

With more than a decade of experience in the fields of nanotechnology and nanotoxicology, I have authored numerous publications, including original research, invited reviews, and book chapters. I am a member of the Society of Toxicology and recently served on the Scientific Advisory Board for the EPA's FIFRA Program. In addition, I serve as Associate Editor for *RSC Toxicology Research* and on the Editorial Board of the journals *Toxicological Sciences*, *Nanotoxicology,* and *NanoImpact*.

I received my PhD in Chemistry in 2005 from Rice University. My dissertation focused on the "nano-bio interface." In 2005, I joined the DuPont Company as Visiting Scientist and aided in the drafting of the DuPont-Environmental Defense Nano Risk Framework.

The 2008 Nano-Effects Working Group at Haskell Laboratory for Environmental Health at the DuPont Company: From left to right: Maria Donner, Christine Glatt, Christie Sayes, Brian Slezak, David Warheit, Carol Carpenter, Xing Han, and Diane Nabb

The 2007 Nanomaterial Aerosolization Team at the DuPont Company: From left to right: Michele Ostraat, Tracy Rissman, Kenneth Reed, and Christie Sayes

Daphne L. Schmidt Life is a journey with many pathways, and I attribute who I am today to a convergence of the paths that were chosen and the serendipitous opportunities that arose unexpectedly. Growing up in an Army family, I had the opportunity to live around the world as my father was stationed in places as far flung as Orleans, France, and the Panama Canal Zone. I truly treasured those experiences of living in completely different cultures, rich with traditions. They planted a seed of curiosity about the world and those that inhabit it that has continued to grow throughout my life.

I attended Virginia Tech, eager to study biology and psychology. While there, I had the opportunity to do some independent research with a psychology professor investigating the role of the thalamus on information processing. This was my first taste of pure scientific research and I found it fascinating. But the time in the labs also helped me realize that I needed to find a career that involved not only science, but also interaction with people. So, after deep consideration, I decided in my junior year to transfer to the Medical College of Virginia to study nursing. It was the perfect fit.

As a newly graduated R.N., I specialized in pediatric nursing and worked in a special care nursery. The nursery attended to any returning newborn ICU infants and served as a step-down unit for the pediatric ICU. It was a high-charged environment that was demanding and exciting, and required the ability to work with multiple health team members and families. I learned a great deal about the complexity of mind and body interactions in health and the power of education to support positive outcomes. During this time, my husband was in medical school and once he graduated the Army moved us away from Virginia and had us crisscrossing between the hills of Texas, the hot sands of the Mojave Desert, and the whirlwind of Washington, D.C. and eventually back to Richmond.

While in Texas for the second time, I decided to make a career switch. We had three small children and although I loved the hospital environment shift work was challenging with on-call schedules and living far from family. I received my K-8/Biology teaching certification from the University of Texas and proceeded on a new career path in science education. As a middle school science teacher in Arlington, VA, I coordinated the school science curriculum, student participation in the JASON Project, and the school science fair, and served as the regional science fair committee member and school liaison. In addition, I successfully applied for several grants to enhance the hands-on learning experiences of our students including a Biotechnology Institute's Genome Project Education grant, an NBC Weathernet weather station grant, and a School Greenhouse Project grant. In 2005, I received my M.Ed. in Administration and Supervision from the University of Virginia and shortly thereafter we moved to Richmond, VA.

Most recently, it was my privilege to serve as the Coordinator of Professional Development and Director of the Summer Regional Governor's School at the MathScience Innovation Center (MSiC), a STEM hub serving 12 school divisions in Central Virginia. I was charged with supporting the MSiC's consortium school division teachers with pedagogically and content-rich STEM education professional development. I was also the project lead for the Center's Nanoscience & Nanotechnology initiative, a program that grew to include a Nano Fellows Institute, short courses, conference model classroom lessons, and student enrichment programs. It was truly an honor to present our work in nanotechnology education to the Triennial Review of the National Nanotechnology Initiative at the National Academies of Science in 2015. Through collaboration with MSiC faculty and adjuncts, college, university, and regional partners, I worked to bring innovative concepts such as nanoscience into the classroom, building a framework of vertical learning experiences for both students and teachers in order to prepare students for future academic and market challenges.

An extension of my position at the MSiC was serving as liaison for our work with Challenger Center Headquarters. After the Challenger disaster in 1986, families of the Challenger astronauts banded together to form a foundation with the mission to continue the education goals of the Challenger mission. Rather than a statue or memorial, the families decided to develop space simulation learning centers that would teach and inspire youth to explore a career in space science and engineering. To date, there are more than 40 Challenger Centers around the world.

Stepping into one of the Challenger simulation centers is like visiting the Johnson Space Center. Students begin their mission in a briefing area, and then move between mission control and a space capsule conducting research experiments with their fellow "astronauts," analyzing it, and making decisions that will guide the mission to a successful conclusion. The MSiC has the only Challenger Center in Virginia. It was thrilling to be able to help bring new missions to our Center exploring everything from climate change and Earth systems, to cutting-edge Mars missions. We also partnered with Challenger headquarters on several grant projects developing virtual space exploration experiences, making an exciting learning opportunity available to more youth across the nation.

I have had the pleasure of working very closely with our regional PBS station (Community Idea Stations) and their Science Matters initiative, both coordinating hands-on activities with my MSiC colleagues to complement programming and serving on the Science Matters Leadership Team. Most recently, I was asked to serve as Chairperson for this diverse group of science and education leaders, committed to supporting the station's goals.

Professional affiliations include the Association for Supervision and Curriculum Development, the National Science Teachers Association, NASA AESP Professional Development Alliance, and the Science Matters Leadership Team. I was previously a James Madison University adjunct faculty member in the office of Outreach and Engagement for the Big Ideas of Nanoscience & Nanotechnology Fellows Institute at the MathScience Innovation Center.

Life is certainly an adventure, with one experience building on another and opportunities sometimes arising in the most unexpected places. I adore the thrill of learning something new and sharing ideas with others. Here's looking to the next great adventure!

Quinn A. Spadola I am Associate Director for Education and Outreach for the NSF-supported National Nanotechnology Coordinated Infrastructure (NNCI), Education and Outreach Coordinator for the Southeastern Nanotechnology Infrastructure Corridor NNCI site, and an Academic Professional in the Institute for Electronics and Nanotechnology at the Georgia Institute of Technology.

Prior to joining Georgia Tech, I was Education and Outreach Coordinator and Technical Advisor to the Director in the National Nanotechnology Coordination Office. I joined that office as an AAAS Science and Technology Policy Fellow in 2014 and joined the professional staff after completing my fellowship. There I worked to educate students, teachers, and the general public about nanotechnology through conferences, contests, networks, videos, and podcasts.

I was inspired to go into education and outreach while working on my Physics PhD, which I earned from Arizona State University in 2008. While at ASU, I became a Center for Nanotechnology in Society-Biodesign fellow. As part of that program, I established a monthly "Science Café" series in the Phoenix area. My goal was to provide an opportunity for interested, nonexpert adults to speak with scientists and engineers and nontechnical experts like lawyers, philosophers, and social scientists. Each conversation focused on a specific research topic and how it affects audience members' lives. Topics included nuclear power and our ability to assess risk, reconciling religious faith and acceptance of evolution, ergonomics in the military, and the future possibility of brain-machine interfaces. While I enjoyed my research, to me, creating opportunities for people to learn about and share their opinions on current research was so much more important and satisfying.

This experience is what led me to attend film school. My PhD gave me a strong scientific background, but not the skills to communicate with a general audience in an effective and entertaining manner. I earned an MFA in Science and Natural History Filmmaking from Montana State University in 2011. I have always been interested in going beyond the "gee whiz, that's cool" aspect of science outreach to include the politics and historical context behind science research. In film school I made videos that integrated science, my experience as a female scientist, and the historical context of gender and STEM.

The next step in my career was to become a AAAS Science and Technology Policy Fellow. Having started in 1973, this competitive fellowship program requires that scientists have a PhD and an interest in and understanding of the societal impacts of science. The purpose of the program is to place scientists and engineers in the federal government in order for them to learn about and contribute to policymaking. This experience allowed me to add an additional dimension to my understanding of the scientific research enterprise. This is also where I became formally involved in nanotechnology education and outreach. Working in this space combined my technical background (my research was on an AFM-based nanopore DNA sequencing method) and my desire to engage multiple audiences with science.

Being able to work directly with students, educators, and the general public in my current position is as satisfying as the Café conversations I helped to start back in graduate school. I hope to continue to inspire everyone from elementary school students to curious adults to learn more about science and engineering.

Nicole F. Steinmetz My inspiration to pursue innovative research reflects my time as a high-level athlete. I began competitive artistic roller figure skating at age 5, and practiced my skills alongside my school work. Competing for the German National Team, I gained top 10 positions in European and World Championships. I won gold at the German Championships (2000), and silver at the European Championships (2002). To achieve the highest rewards in internationally competitive sports one has to be creative and innovative, and extremely disciplined, and must also have the courage to take risks. Through training and competing in skating I achieved self-discipline, focus, dedication, unlimited enthusiasm, and a love of competition at the highest levels. My time as a high-level athlete had a great

 impact on my scientific career. I learned to take risks and to be bold and daring when learning new and difficult jumps, and to always stand up and try again until I succeed.

My interest in bio-nanotechnology was sparked when, as an undergraduate student, I attended a lecture on the molecular farming of antibodies in plants using expression cassettes derived from plant viruses. I was immediately fascinated by the idea that we can reengineer nature's nanocarriers—plant viruses—and use them to target applications in human and plant health. I have studied plant virus nanotechnology since my undergraduate days, and during each step of my scientific training I focused on a different aspect of technology development. In my early work I focused on understanding how plant viruses interact with their natural hosts and I optimized molecular farming protocols for the production of pharmaceuticals in planta (Master's Thesis, 2003–2004, RWTH Aachen' Germany). During my PhD research as a Marie Curie Early Stage Training Fellow (2004–2007, John Innes Centre, Norwich, UK), I studied the development and nanofabrication of electronic virus-based arrays for applications in sensor and chip technology. As a postdoctoral fellow (2007–2010, the Scripps Research Institute, La Jolla, CA, USA) funded by an American Heart Association Fellowship and a National Institutes of Health K99/R00 award, I then turned toward medical applications and I established the design principles that allow plant virus nanoparticles to be used for in vivo targeting in mouse models of human disease.

This launched my career as an independent researcher. I started my faculty position at Case Western Reserve University School of Medicine and joined the Biomedical Engineering Department, where I was promoted through the ranks of Assistant Professor (10/2010), Associate Professor (07/2016), and Full Professor (01/2018). I held the George J. Picha Designated Professorship in Biomaterials and served as Director of the Center for Bio-Nanotechnology. In 2018, I joined the University of California, San Diego, as Professor of Nanoengineering. My research efforts have been recognized by many awards. For example, I was elected fellow of the American Institute of Medical and Biological Engineering (AIMBE) in 2017, and I was recognized as a 2016 American Cancer Research Scholar, a 2015 Young Innovator in Cellular and Molecular Bioengineering, and a 2014 Cleveland Crain's 40 under 40 Awardee. I have authored more than 150 peer-reviewed journal articles, reviews, book chapters, and patents. I have also authored and edited books on virus-based nanotechnology. We have been fortunate to land many research grants to support our research program. My research is funded through grants from the NIH and NSF as well as the Susan G. Komen Foundation, American Cancer Society, and American Heart Association. Over the past 8 years, I have attracted grants as principal or co-principal investigator exceeding $18 million.

Our research program focuses on the development and testing of plant virus nanoparticles with applications in medical imaging, drug delivery, and immunotherapy. I enjoy leading an interdisciplinary research program and team, bringing together an international group of postdoctoral fellows and students from the biomedical

engineering, chemistry, biology, botany, and immunology disciplines. My first PhD students, who graduated in 2016, are the co-authors of the chapter we have contributed to this volume. Amy Wen is now a postdoctoral fellow at the Wyss Institute at Harvard University, and Karin Lee is a postdoctoral fellow at the National Cancer Institute. Both were terrific students and colleagues in my laboratory, and set a high bar for other students to follow. I am sure they will have an immense impact as women leaders in the field of nanotechnology.

In addition to training the next generation of scientists and engineers in bio-nanotechnology, I am eager to make nanotechnology accessible to the general public. Toward this goal, we have taken an interdisciplinary approach that mixes STEM with the performing arts. In a collaborative effort between myself and Knight & Brinegar, a retro-forward musical-writing team, we developed "The Nanoman"—a project that bridges the fields of nanotechnology, gaming and graphic design, and theatre, to capture the concepts of drug delivery and cancer nanotechnology. Across various media platforms, our tiny superhero "The Nanoman" is on an important mission: "*Go, go, Nanoman: find and kill that tumor man!*" In video clips, music videos, video games, and live incarnations, we aim to make science accessible, entertaining, and enlightening. We use story, music, and interactivity to explain the challenges of current cancer treatments as well as the engineering principles that can be applied to enhance cancer therapy, with the ultimate goal of improving patient survival.

The fun on the computer screen, tackling cancer with superpowers, is the daily reality in our laboratory. Although we are still early in the development and validation stages, we have successfully demonstrated our virus-based cancer immunotherapies in various mouse models and even in the treatment of companion dogs diagnosed with late-stage melanoma, and we are driven and focused to move this technology toward clinical testing. Nevertheless, we have just scratched the surface of this technology and many more discoveries await us!

Evelyn N. Wang I grew up in an academic family where both of my parents received their PhDs from MIT—my mother in chemistry and my father in electrical engineering. I had two older brothers that excelled in math and science. While I also enjoyed learning, I spent much of my childhood playing the piano and violin. It was a way to help differentiate me from my brothers and it allowed me to express myself in another way, given that I was also very shy. I had the opportunity to perform a lot and also play in orchestras and chamber music, where I was able to form many strong friendships. It also helped me learn how to be a leader as I was the concert mistress for my orchestra for 3 years where we travelled to perform in many places around the world.

While I loved music, I knew that I would not pursue it as a profession. I chose mechanical engineering as an undergraduate at MIT because I thought that it could combine the elements of math, science, and artistry that I enjoyed. The breadth of the major allowed me to be exposed to so many things. Ultimately, I decided to pursue graduate school at Stanford University where I worked at the intersection of heat transfer and microelectromechanical systems (MEMS). Specifically, my research focused on new two-phase heat dissipation strategies using silicon-based MEMS devices for thermal management of integrated circuits. Graduate school was a wonderful time. While there were many moments of uncertainty, it was also a time that allowed for an intense focus on research, which I enjoyed. I was also fortunate to have tremendous advisors that gave me the flexibility to pursue my interests and yet gave me guidance when I needed it most.

After I received my PhD, I joined Bell Laboratories, Alcatel-Lucent, as a post-doctoral associate for 1 year. It was an incredibly eye-opening experience as I became exposed to another world of nanoengineered surfaces. I was fascinated by the ability to use such approaches to have fine control of interfacial phenomena. My exposure to this area opened up many new possibilities as I started my career as a faculty member in the Mechanical Engineering Department at MIT. While I never anticipated being back at MIT, it felt like home when I returned. I had, again, tremendous support from my colleagues, the department, and MIT. I started my research program focused on using nanoengineered surfaces to enhance heat and mass transport processes. My work has since expanded to applications beyond thermal management, including energy conversion and water desalination, among others. The opportunities as a professor have been tremendous. It has given me the flexibility to explore curiosities and pursue research avenues that I find meaningful. I have been able to interact and learn from so many talented colleagues, students, and postdocs. I can truly say that it has been my dream job, and I have been very fortunate to have the opportunities that have guided me towards this path.

Amy M. Wen My journey that led me to research in nanotechnology was directed by fortuitous events and fantastic mentors. My primary passion since I was young has been in problem-solving. I highly enjoy the "Aha!" moments that come when finally solving particularly challenging puzzles, especially ones that require either tricky and unusual or simple and elegant approaches. All throughout my schooling, my family and teachers have all wholeheartedly encouraged my interest. Moreover, they went above and beyond to provide additional resources for my classmates and me, such as using their spare time to conduct sessions on interesting topics such as game theory. Their enthusiasm fueled my own and led to exciting opportunities, including a chance to represent my state of

North Carolina at the American Regions Mathematics League competition. It was a gratifying experience to participate as part of a tremendously talented team and also to be able to contribute one of the top scores that led to our success against over 100 teams both nationally and internationally.

While I enjoyed the challenge of participating in competitions, I also wanted to do more and utilize my problem-solving skills to reduce and prevent human suffering from disease. The solution was obvious when I became aware that a university just a few blocks away from my high school had a nationally recognized program in biomedical engineering. Biomedical engineering was the perfect blend of the problem-solving aspects of engineering being used to address the scientific challenges in healthcare and medicine. As an undergraduate at Duke University, I had the opportunity to participate in the Pratt Research Fellows Program, which matched labs interested in training interns with students interested in particular labs' research. I was drawn to Dr. Ashutosh Chilkoti's lab, where bioinspired elastin-like polypeptides that self-assemble to form micelles were being used for drug delivery. Under the guidance of a graduate student mentor, (now Dr.) Wafa Hassouneh, I used molecular cloning to introduce calcium-sensitive regions so that the micelles could be triggered to release drug cargo specifically within cells due to lower intracellular calcium levels. This was my first exposure to hands-on research in nanotechnology and really sparked my curiosity and drive to continue working in a similar area in graduate school.

I was captivated by Dr. Nicole Steinmetz's lab at Case Western Reserve University because I thought her research using plant viral nanoparticles was a unique, largely unexplored approach with some interesting advantages I wanted to explore, such as their monodispersity and ease of manufacture. I would also be able to continue with the theme of applying bioinspired nanoparticle platforms for therapeutic applications. While Nicole had an extraordinary record of accomplishing a lot of amazing and high-quality science in the past, there was some hesitation because she was a new professor and the lab was not yet established. One of the deciding factors that led me to select her lab anyway was the invaluable advice Wafa gave me, which was that at the end of the day it is excitement over your research that gives you the motivation to work harder, not necessarily the seniority of your professor or the location of the lab. I was extremely fortunate that not only was the research stimulating, but Nicole also proved to be an exceptional mentor. Beyond providing research guidance, she also served as an example of how to get a new lab running, offered many opportunities for collaborations, provided a supportive environment of fellow researchers (including our co-author Dr. Karin Lee and proofreader of this biography Dr. Neetu Gulati), encouraged us to present our work at conferences, advised us on writing fellowships and grants, and overall strengthened my passion for research. Under her mentorship, I was able to work on a number of different projects studying aspects of nanoparticle design, shape, payloads, and targeting for the development of virus-based nanoparticles for both drug delivery and imaging applications, with the support of training grants and predoctoral fellowships from the American Heart Association (AHA) and the National Institutes of Health (NIH).

When searching for postdoctoral positions, I sought to expand my skill set to supplement my experience with nanotechnology. In particular, I wanted to tackle the challenge of streamlining the translation of discoveries made in the lab into the clinic. While my graduate research provided a lot of experience with in vitro cancer models and in vivo mouse models, it was clear that a scientific gap exists in translating the results from such experiments into humans. Although there are established animal models that are known to be more predictive of clinical efficacy, the results are often still inaccurate due to intrinsic differences between humans and other species. Therefore, I was keen to join Dr. Don Ingber's group at the Wyss Institute for Biologically Inspired Engineering at Harvard University for my postdoctoral training in order to learn methodologies that could more accurately recapitulate human biology. As part of my training in his lab, I have recently been awarded an NIH postdoctoral fellowship to investigate how different environmental factors affect the development of pulmonary fibrosis in a microfluidic lung-on-a-chip model and to utilize insights gained from the model for discovery of novel therapeutic approaches. The biomimetic microsystem more closely recapitulates human physiology by integrating fundamental mechanical cues from the lung microenvironment, such as tissue-tissue interfaces, fluid flow, and breathing motions, and our results can be used to complement and minimize the use of animal models when testing for safety and efficacy.

In the future, my goal is to combine my background and experiences in nanotechnology and disease modeling to advance understanding and develop strategies for the diagnosis and treatment of diseases. I am extremely grateful for the friendship and support of an excellent group of friends, colleagues, and mentors, and I aspire to reciprocate in kind by providing an encouraging and supportive environment wherever I go. I have come a long way from my roots in unraveling relatively straightforward puzzles with known solutions to now tackling more complex and open-ended challenges where solutions are yet to be discovered, and I am even more excited and passionate now to do my part to solve these problems.

 **Mona Zebarjadi** My grandfather was living in a small rural area of Iran, called Ferdows. The world "Ferdows," means heaven. Perhaps because this was the only relatively green spot in a vast desert. Half of my family members still live there. My grandfather used to think that girls should stay home and therefore did not allow his girls to attend school. An earthquake apparently changed the faith of the family. The family lost their house and chose to move to the city to rebuild their path of life. After moving to the city, my grandfather was convinced that he should let his girls go to school. He did so for all his girls, but it was too late for his oldest daughter, my mom.

As a kid, I used to draw a lot. I never felt that I am talented in that regard, but I was enjoying the mindfulness of it. Our neighbor, seeing my passion, introduced me and family to one of the masters of oil painting in the city, Mr. Shahbazi. I went to his class for a semester before my family realized that we cannot afford the tuition. Mr. Shahbazi though told my family that he will teach me free of charge. I became a long-time student of him and only learned about the arrangement when I was at college. Mr. Shahbazi was a great person and I soon took him as my role model. It was amazing to watch him paint. As soon as he was holding his brush, he was in a different world, not hearing anything else. I always wished I had such a passion for something, anything. I was determined to find my own brush.

At school, I knew I was good in math and physics, but I also knew that was just relative to other students, so I did not think of myself as being especially talented in that regard either. I did not have a passion for math as I looked at it as only a tool, but I always enjoyed physics concepts so I chose to study physics. I was accepted in Sharif University, the top engineering and science school of Iran, allowing me to continue my education free of charge. So I left my family and moved to the capital. It was a very difficult experience. Back in high school, I did not have to study outside of the classroom but it was very different in college. Sharif University had a strange environment. Apparently, some of the students would enter thinking too high of themselves and rude to the faculties. So the educational system was designed to make sure to communicate well that you are not as good as you think. Perhaps that was necessary for some, but for most others such as myself with very little self-confidence, it was just overwhelming. That policy along with many family problems caused a deep depression that persisted in my first 2 years of studying. I was constantly seeking medical advice but because of the side effects of the medicines, I did not allow myself to be on them for long. I am not sure why I stayed, perhaps because I did not want to go back to my family considering that the only thing waiting for me back home was an unwanted marriage. There was no point in switching my field either as there was nothing else that I liked more than physics. I started painting again which helped me significantly with my depression. My grades also got better over time and I finished my bachelor with a GPA well above the department average despite my shaky start.

At that point, most my friends were applying to get acceptance from schools in other countries because of the unhealthy economy as well as social restrictions in Iran. I soon realized that leaving the country unmarried was just impossible with my parents. So I decided to enter the master program. A friend who is now a faculty in the United States suggested that if I work with a professor in the United States for my master thesis, I might be able to build a bridge. I surely did not share all my problems with her, but I realized that if I work with this originally Iranian professor, Prof. Ali Shakouri, for 2 years and if my dad feels secure enough, he might let me go. So I accepted the project, thinking I will do it no matter how hard it is, and I actually did it. I reached to a point that my co-advisor in Iran suggested me to go to Turkey as an exchange student to learn Monte Carlo technique (which I needed for the project) from the experts.

Growing up, we never had a family trip to anywhere except for northern part of Iran where my dad had a hotel deal offered by his bank. So this was my first trip outside of the country. I spent 2 months in Turkey under the supervision of Prof. Bulutay. Nothing in the world was better than that 2 months. I quickly realized that I can take care of myself! Prof. Bulutay assured me that I am safe in the campus. So I dared taking lonely walks in the middle of the night. I was working hard and enjoying myself at the same time. I was working sometimes until 2 am and was walking home without any fears. I was going to all sorts of concerts, ballets, and classic music thinking I will never have the chance to do anything like that again in my life. At the end of the 2 months, I had a lot of results and all three advisors were happy with the outcome of my work. I think that was the first time that I thought it is possible for me to stay in the field. It was the first time that I truly enjoyed what I was doing and my achievement. This was the brush that I was looking for.

Many ups and downs have happened after that. I moved a lot and I changed my school, and my field a lot. But once you have your brush, everything is much easier. The rest is just catching up with the ever-changing flow of life.

Many would tell you that you should dream big. But what if you are in an environment that you do not even know what dreaming big means? What if you cannot find your brush? What if you do not even know where to start? Being a faculty member at University of Virginia today, having two lovely boys and a great scientist as my husband, I feel very lucky. I grew up in an environment where I did not even know what it means to be a faculty, a researcher, or an engineer. I was unable to answer when people asked me what you want to be when you grow up. I was too shy to ask many questions. All I knew was what I did not want to be. I never dreamed big! All I did in my life was to try really hard to choose the best option available to me at the moment and to be grateful for it. I did not have much guidance growing up and perhaps that is one of the reasons I am extremely grateful to the mentors that I found later on in my life. The world is a big one and a small one at the same time. Some would know where they want to be early in their life and some finds it only at a later age. No matter where you are from, if you do choose the path that you think is right for you and not the ones that others try to impose to you, you will find the mentors that you truly need.

Yajing Zhao I want to make my academic path a unique adventure. Upon middle school graduation, I made one of the biggest decisions in my life—I attended the Special Class for the Gifted Young of China (SCGYC) program at Xi'an Jiaotong University, which allowed me to experience college life starting at the age of 15. The SCGYC program enabled me to largely bypass the traditional exam-centric curricula in normal Chinese high schools, and instead gave me significant room to explore knowledge of interest, such as discussing advanced mathematics problems with classmates, and conducting scientific experiments in university laboratories at an early age. In retrospect, my experience at SCGYC not only let me make treasured friends with talented peers, but also preserved and advanced some of my characteristics such as curiosity and fearlessness.

Choosing my undergrad major was another important decision I made that will influence my life. Math and biology have always been my interests since I was a little kid. However, I ended up choosing energy and power engineering as my undergrad major due to two considerations. First, my undergrad school is most well known for its engineering programs, while math and biology majors are not as highly regarded. Second, energy seemed to be the most attractive field to me among all the engineering programs. At that point, I did not realize how much my choice of undergrad major would influence my career path. Looking back, it was the moment when I journeyed down the road of studying thermal engineering. All the professional training I received, and all the academic mentors and peers I interacted with during my undergrad years, gradually cultivated in me my expertise in heat transfer, fluid mechanics, and thermodynamics.

I started my undergrad research by working on thermo-economic analysis of geothermal power plants. Geothermal energy as a source of sustainable energy is expected to make an increasing contribution to energy supply in the near future. However, there has been a trade-off in the energy efficiency of geothermal power plants and their economic benefits. Thus, I spent months developing a model to analyze and optimize both the thermodynamic and economic performance of a geothermal power plant system. My undergrad research experiences kindled my passion for advancing the frontiers of energy science, and made me believe that I am competent to explore the unknown in this field. In this case, it became a natural progression for me to pursue graduate study where I would be able to reach my full potential.

My life at the Massachusetts Institute of Technology has been a magical journey. I got to know and became greatly impressed by Professor Evelyn Wang's work through MIT News when I was a sophomore. Getting into MIT and joining Evelyn's lab is like fulfilling a dream. At graduate school, I became interested in enhancing condensation heat transfer via micro/nanoengineered surface designs. Traditional condenser surfaces, such as those used in power plants, usually form a millimeter-thick condensate film due to the high surface energy of common condenser materials. This condensate film poses an intrinsic barrier to heat transfer and therefore limits the energy efficiency of the overall system. Recent advances in nanotechnology and fabrication enable us to design and fabricate micro/nanostructured surfaces whereby we can manipulate the wetting behavior of the condenser surface and improve the condensation heat transfer performance. Previously I have worked on slippery liquid-infused porous surfaces (SLIPS) for condensation of low-surface-tension fluids. My current work focuses on utilizing capillary pressure generated by hierarchical porous surfaces to enhance condensation of water. With a deep appreciation for nature and wildlife, I am also passionate about bio-inspired surface designs for various energy-related applications.

Yangying Zhu With a curiosity for science and architecture, I attended Tsinghua University in China for my undergraduate study majoring in mechanical engineering, where I studied heat transfer and energy-efficient heating and cooling technologies for buildings. Through class and research intern opportunities, I realized that understanding the fundamentals at the small scale is a key to improving the macroscopic system-level performance. I decided to pursue a PhD in mechanical engineering focusing on microscale thermal and fluid transport. This journey completely opened my eyes to the beauty and power of micro/nanotechnology. I developed skin-inspired surfaces with microscopic "hair" arrays that can dynamically tilt in a magnetic field to manipulate fluid and light. I also applied microstructured surfaces to enhance heat transfer for cooling high-power electronics. It was fascinating to learn that small structures can make a big impact. I also enjoyed being a researcher and loved brainstorming with colleagues and mentors to find creative solutions. What excites me most is the process of searching the answers for the unknown and making new discoveries.

As a scientist, I believe that energy storage and clean energy technologies are crucial for energy and environmental sustainability. After I graduated from MIT, I worked as a postdoc in the Materials Science Department at Stanford exploring heat and mass transport problems in batteries and electro-catalysis systems. Even though energy storage and conversion are new fields to me, the research skills I obtained in my PhD training allowed me to define an interesting problem to investigate and learn related knowledges along the way. I recognized that many future opportunities lie in the interdisciplinary area. As I move forward to start my independent research career, I hope to build a program that intersects the fields of thermo-fluid engineering and materials science for more efficient thermal management and sustainable energy solutions.

I am fortunate that along this journey, I am accompanied by many female peers and mentors who are my role models. Inspired by them, I volunteered to mentor high school students from Thayer Academy on their science internship at MIT, during which I introduced to them basics of nanotechnology, including fabrication techniques and imaging tools to visualize microscopic material structures. It felt very special for me to see their curiosity and excitement because my own interest in engineering originated from high school when a few MIT undergraduate students taught us over a summer course to build an airplane model. In my future career, I hope to share my passion with more young women, and to encourage them to explore the beauty of science and engineering.

Chapter 1
Introduction to Nanotechnology

Lisa E. Friedersdorf and Quinn A. Spadola

Introduction

Nanotechnology is the study, manipulation, and application of matter at the nanoscale, where the nanoscale refers to lengths of less than a hundred nanometers (nm). To put this into perspective, a typical sheet of paper is 100,000 nm thick and a human hair is on the order of 80,000 nm in diameter, so this is almost unbelievably small. Scientists, engineers, and developers are interested in working at these size scales because materials at the nanoscale can behave differently than the same material at the bulk or macroscale. Depending on the material, at very small size scales, they may become stronger, more ductile, damage resistant, conductive, or chemically reactive. Even properties like color depend on size at the nanoscale. While nanotechnology remains an active area of research in almost all areas of science, engineering, and medicine, a vast array of applications and products have emerged that take advantage of the novel properties of nanomaterials.

L. E. Friedersdorf (✉)
Department of Materials Science and Engineering, University of Virginia,
Charlottesville, VA, USA
e-mail: lef2p@virgina.edu

Q. A. Spadola
Institute for Electronics and Nanotechnology, Georgia Institute of Technology,
Atlanta, GA, USA
e-mail: quinn.spadola@ien.gatech.edu

© Springer Nature Switzerland AG 2020
P. M. Norris, L. E. Friedersdorf (eds.), *Women in Nanotechnology*, Women in
Engineering and Science, https://doi.org/10.1007/978-3-030-19951-7_1

Why Size Matters

Some of the observed changes in material behavior at the nanoscale are due in part to the increased role of surface atoms. In bulk materials, the number of surface atoms is a small fraction of the total number of atoms and average material properties are based on the atoms in the interior of the material. For very small materials, however, the properties of the surface atoms dominate and can change the chemical behavior of the material. For example, while bulk gold is considered inert, it becomes reactive or even catalytic at the nanoscale. Figure 1.1 uses two-dimensional arrays to illustrate this point. In this figure, the array on the left is made up of nine "atoms," of which eight are on the outside or are "surface" atoms. The array or "particle" on the right is made up of 81 "atoms," of which 33 are on the outside. Therefore, in this simple illustration, 89% of the atoms of the small "particle" reside on the surface, while only 41% of the atoms that make up the slightly larger "particle" on the right reside on the surface.

In addition to its reactivity, size also impacts the color of gold. At the macroscale, gold is, of course, gold or yellowish, but below a certain size gold can appear red or blue. This effect illustrates the importance of quantum effects at the nanoscale. In noble metal nanoparticles, free electrons can collectively resonate in response to incident light. As a result, smaller metal nanoparticles absorb shorter wavelengths, the blues, and reflect back the longer wavelengths, giving them a reddish color. As the gold nanoparticles get bigger, the wavelength absorption shifts resulting in more red light being absorbed while wavelengths at the blue/purple end of the visible light spectrum are reflected. When small nanoparticles are made from semiconducting materials, they are typically called quantum dots. Quantum dots illustrate quantum confinement where either the electrons or the holes are confined resulting in

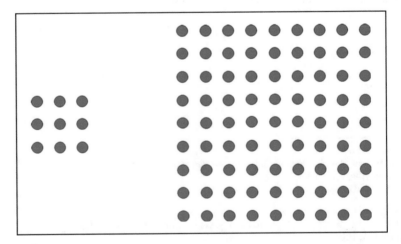

Fig. 1.1 Size matters. Two-dimensional arrays illustrating influence of size on number of surface atoms

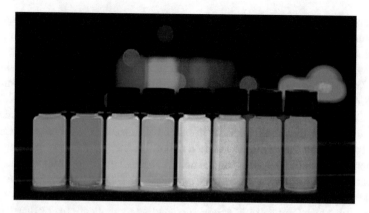

Fig. 1.2 Quantum dots. The image shows ZnCdSeS alloyed quantum dots from PlasmaChem. The emission wavelength of a quantum dot can be tuned by changing its diameter. This image is used under the Creative Commons Attribution-Share Alike 3.0 Unported license

discrete energy states for the overall particle. This is sometimes referred to as an artificial atom. As Fig. 1.2 shows, larger quantum dots emit longer wavelengths of light when excited, giving off red or orange colors. Smaller dots emit shorter wavelengths which results in blue and green colors. Working at the nanoscale allows scientists and engineers to tune the optical properties of nanoparticles by adjusting their size and shape (Kreibig and Vollmer 1995; Woggon 1997).

Reactivity and color are just two examples of how material properties are influenced by size. Atomic arrangement and bonding also influence material properties. To illustrate this point, consider two well-known bulk materials, graphite and diamond. Graphite and diamond are two of the allotropes of carbon. Allotropes are different physical forms of material made from the same element, in this case carbon. In graphite, the carbon atoms are arranged hexagonally in sheets, and the resulting material is black and slippery. Within the sheets, each carbon atom is bound to three others and these bonds are very strong. The bonds between sheets are weak, however, which is why material is left behind when a pencil is traced across a sheet of paper. In diamond, the carbon atoms are arranged tetrahedrally; it is transparent and very hard and strong. The carbon atoms in diamond are each bound to four others in a three-dimensional arrangement called the diamond cubic structure. While both graphite and diamond are made entirely of carbon, their properties are very different because of how their atoms are arranged and the bonding between them.

There are several other allotropes of carbon that have extraordinary properties and the discovery of these carbonaceous nanomaterials fueled much of the early enthusiasm about nanotechnology. Hollow molecules of carbon called fullerenes were discovered in 1985, the most famous of which is C_{60}, the Buckminsterfullerene, after Richard Buckminster Fuller, the architect renowned for geodesic domes (Kroto et al. 1985). Nicknamed buckyball, this molecule has a truncated icosahedron structure, like a soccer ball, with 60 carbon atoms. The identification of another form, carbon nanotubes (CNTs), was published in 1991 (Iijima 1991). CNTs are like a single layer of graphite with the carbon atoms arranged hexagonally, rolled in a tube with a diameter on the order of nanometers. Single-walled CNTs refer to a single tube, where multi-walled CNTs refer to a tube, in a tube, in a tube, depending on the number of walls. When CNTs were characterized, it was discovered that they were the strongest (100 times stronger than steel) and stiffest materials known to exist (Yu et al. 2000). Since CNTs are also low in density, excitement grew about applications, such as an elevator to space, that could be made possible with these strong, lightweight materials. Depending on their chirality, carbon nanotubes can exhibit very high electrical conductivity (metallic behavior) or can behave more like a semiconductor, leading to applications in electronics. The optical properties of carbon nanotubes, including tunable absorption, have application in areas of optics and photonics. Carbon nanotubes are also excellent thermal conductors leading to applications for thermal management. Although there remains much work to be done to fully understand and produce CNTs with the desired properties, they are already being used in a wide variety of applications from strong lightweight composites in automobiles and sporting goods, light-filtering window films, enhanced gas sensors, and radiation-hardened electronics in space.

Allotropes of carbon continue to be discovered and developed, including graphene, a single layer of graphite, referred to as the wonder material and leading to a Nobel Prize in 2010 (Novoselov et al. 2004; Royal Swedish Academy of Sciences 2010). The discovery of graphene was important to nanotechnology not only because of its exceptional properties (strong, flexible, electrically and thermally conductive, impermeable, etc.) but also because it opened up an entire new field of two-dimensional (2D) materials.

One of the most influential scientists in the area of carbonaceous nanomaterials was Mildred Dresselhaus (American Physical Society 2018). Known as the "Queen of Carbon Science," Dresselhaus was a solid-state physicist who made critical advances in the understanding of the thermal and electrical properties of buckyballs, carbon nanotubes, and graphene, often developing the methods to characterize and measure these properties. She is also credited with introducing the field of low-dimensional thermoelectricity for the conversion of a temperature gradient into electrical energy.

Materials by Design

Building at the nanoscale is often broken into two categories, top-down or bottom-up. An example of top-down approaches is lithography techniques that allow engineers to carve away thin layers of material to create nanoscale features. This is common in the fabrication of nanoelectronics. Bottom-up methods include self-assembly in which scientists rely on understanding molecular interactions to design a construct that will form when they combine the pieces. DNA origami is a classic example of self-assembly in which scientists mix together strands of synthesized

DNA which will hydrogen bond into predicted shapes. At the nanoscale different forces become dominant. Gravity strongly influences how objects behave at the macroscale. Shrink a material down to the nanoscale and the intermolecular interactions or van der Waals forces play a much greater role in how something behaves. These differences can be exploited to make materials by design. The new tools available to manipulate and build at the nanoscale have led engineers to create smart materials that can both sense and respond to their environments with applications in everything from bridge maintenance to dental braces.

There are certain expectations when dealing with materials. One assumes that strong materials will be heavy, and light materials will be weak. Julia Greer, however, disagrees. Greer is proving that engineers no longer need to be bound by these traditional material properties (Greer 2018). By harnessing the novel properties that exist at the nanoscale, entirely new classes of materials can be made with never before obtained characteristics. She focuses on using nanoscale building blocks, nano-pillars, -pipes, and -tubes, for example, to construct larger materials. She has developed a method to combine nanomaterials with their size-dependent properties and hierarchical architectural elements to create entirely new combinations of properties, such as materials that are strong as steel and light as a feather. To illustrate the effect in a ceramic material, for example, she has demonstrated that a highly porous ceramic material with wall thickness of 50 nm fails and crumbles under compression, just as a brittle ceramic would be expected to do. By reducing the wall thickness to 10 nm, however, the structure does not fail under the compressive stress, and actually springs back when the load is removed exhibiting elastic behavior that would be expected in a metal (Meza et al. 2014). The reduction in wall thickness in this example puts the ceramic material into the regime where novel behavior begins, and the ceramic becomes tougher. Using the two-photon lithography method her group developed, they create structures with unique properties, such as exceptionally strong but very lightweight. There are many applications that can take advantage of the unique attributes of these structures. Examples include battery cathodes, photonic crystals, and three-dimensional scaffolds for bone regrowth. The most significant roadblock to large-scale application of this technique is scale-up.

One challenge of nanotechnology is understanding interactions at the nanoscale in order to purposefully take advantage of these interactions to design new nanomaterials. Sharon Glotzer is identifying the rules of interaction and developing the computational tools needed to achieve materials by design (Glotzer 2018). Her work is not limited to naturally occurring structures or arrangements of atoms. These future smart materials will likely possess coupled properties not typically seen together in a single material and ideally will be programmable, reconfigurable, and adaptable to environmental cues. Glotzer has shown that computational materials science is becoming a reality. Her models examine how hard nanoscale particles spontaneously organize in the absence of interparticle forces, solely due to the entropy in the system. By understanding the behavior of nanoparticles, her modeling work has opened up the design of completely new materials. Using this approach, in over less than a decade, her lab was able to go from predicting the assembly for a specific type of nanoparticle to predicting phase diagrams for entire families of nanoparticles. With increased understanding, improved models, and expanding computational power, the Glotzer team identified trends and are now able to design optimized nanoparticles for targeted assemblies.

Inspired by Nature

The nanoscale is where much of nature works. For example, the diameter of double-stranded DNA is 2.5 nm and ribosomes, which translate mRNA into proteins, also fall within the nanoscale. Plants and animals use nanoscale structures to fight infection, repel water, cut down on reflected light, and create vibrant colors. Scientists, motivated by what they see in nature, are mimicking these structures for use on medical implants, solar cells, and adjustable camouflage. For example, the hierarchical nanostructures observed on lotus and hosta leaves are being replicated to impart superhydrophobicity on surfaces to make self-cleaning windows and wind turbine blades that not only stay clean, but do not ice up in harsh environments. This technology has been commercialized in coatings that can be sprayed on everything from concrete to work boots and gloves, or even mobile devices that can be rinsed under a faucet. Nanostructures can even increase water availability. The structure on the back of the Namib beetle has been replicated by a company to harvest water from fog (NBD Nano 2018).

Structural color, such as the vibrant blue of the Morpho butterfly, has inspired the use of texture not only for color, but also to improve performance of photovoltaics. Entrepreneur Marcie Black's first company, Bandgap Engineering, developed nano-engineered solar cells, also known as black silicon solar cells. By reducing energy loss through reflection, the nanotexture increases the efficiency of the solar cells and enables more power to be produced from the same exposed area. Using a low-cost process called metal-assisted chemical etching reduces the cost to produce the silicon solar cells. This technology is now used to make most of the multicrystalline silicon solar cells on the market. Black's current company, Advanced Silicon Group (ASG), is using the same technology to develop nanotextures for biosensing (Advanced Silicon Group 2018). ASG's silicon nanowire biosensors can measure the distinct concentrations of many different proteins in a solution. One target application of this low-cost and quantitative sensor platform is for the tracking of progress of lung cancer treatment to improve patient outcome and lower health costs.

Inspired by the way abalone builds its shell, Angela Belcher engineers custom materials using biology as her factory with a technique to manipulate viruses to build nanostructures (Belcher 2012). Nature produces impressive materials through processes enabled by proteins coded at the genetic level. Naturally occurring biologically formed structures include complex SiO_2 nanostructures formed by diatoms and small single-domain Fe_2O_3 magnets used for navigation by magnetotactic bacterium. These microorganisms have a DNA sequence that codes the protein sequence providing the instructions or the "blueprint" to build the nanostructure. Belcher wondered if simple organisms such as bacteria and viruses could be convinced to work with the rest of the periodic table and if a DNA sequence could be determined to code the protein sequence corresponding to a desired structure. Using this technique, she engineered viruses to express the ability to grow nanowires and self-assemble through a process of directed evolution (Nam et al. 2006). She has also engineered viruses to pick up carbon nanotubes and grow TiO_2 around them for solar cells. Belcher's exquisite structures are built under ambient conditions in a process using nontoxic chemicals with the release of no toxic chemicals into the environment. This research has led to the formation of two companies. Cambrios Advanced Materials uses Belcher's technique to synthesize inorganic materials from soluble precursors and assemble these materials into functional nanostructures. Siluria Technology focuses on the conversion of natural gas into transportation fuels and commodity chemicals enabled by chemical catalysts that eliminate the need for high-temperature and -pressure processes. Looking to nature, scientists and engineers like Belcher are now able to purposefully build and manipulate at the nanoscale.

Small Science Solving Big Problems

The potential to one day detect and treat cancer at the cellular level created a compelling vision for nanotechnology in medicine. Paula Hammond has made significant advances in the fight against cancer (Hammond 2015). Using molecular engineering, she has developed a way to treat even the most aggressive forms of cancer. She has likened cancer to a supervillain that is "clever, adaptable, and very good at staying alive." The villain's superpower stems from genetic mutations that enable new modes of survival allowing it to live through even the best available chemotherapy treatments. Using molecular engineering, there are ways, however, to turn off a gene. Molecules that are short sequences of genetic code, known as small interfering ribonucleic acid (siRNA), can interfere with gene expression, effectively directing a cell to block a specific gene. Therefore, by treating a cancer cell with the gene-blocking siRNA to prevent the survival genes from being expressed, and then dosing it with a chemotherapy drug, it could be destroyed. There are challenges, however. While siRNA works well within a cell, it degrades within seconds when exposed to enzymes that exist in the bloodstream or other tissues. Hammond's super weapon against cancer is a nanoparticle with a core that contains the chemotherapy drug. A very thin layer of siRNA is wrapped around this core. To protect the negatively charged siRNA, a positively charged polymer is used to cover and protect it in the bloodstream. To successfully reach the tumor the particle needs to get past the body's immune defense system which identifies and destroys foreign objects in the bloodstream. So, one more negatively charged layer is added to the particle that serves two purposes. The layer is a highly hydrated polysaccharide (that naturally resides in the body) surrounded by a cloud of water molecules that acts as an invisibility cloak. This layer enables the particle to travel through the bloodstream long and far enough to reach the tumor without being eliminated by the immune system. The molecules in the outer layer also allow the particle to bind to the tumor cell.

Once bound, the nanoparticle is taken up by the cancer cell and can deploy its payload. By modifying the layers, this technique can be personalized. As doctors and researchers identify new mutations, additional layers of siRNA can be added to silence these genes. Also, the type of drug at the core can be varied depending on the type of cancer or developments in more effective chemotherapy treatments.

Using nanotechnology to combat cancer is also the focus of Michelle Bradbury. The goal of her work with translational silica nanomaterials is to help doctors detect, diagnose, and treat cancer (Bradbury 2018). The ability to make cancer, in particular diseased lymph nodes, glow by shining light on silica nanoparticles will revolutionize surgery. Doctors will be able to map diseased sites in the body and surgeons will be able to actually see which cells are cancerous and remove them while leaving behind healthy tissue. The nanoparticles accumulate on the cancer cells because of peptides on the outside of the particle which selectively bind with tumor cells. The nanoscale size of the particles allows patients to eliminate nanomaterial that does not bind to the targeted cancer.

While considerable progress has been made toward the goal of early cancer treatment, the areas impacted by nanomedicine are much broader than cancer alone. The large surface area-to-volume ratio of nanomaterials along with the ability to functionalize them is ideal for sensors. Nanomaterials have been used in sensing applications, such as over-the-counter pregnancy tests, for many years. Perena Gouma develops nanostructured sensing elements that detect specific biomarkers (Gouma 2018). Her "electronic nose" can be used to detect markers of disease in people's breath and monitor infection or diabetes, for example. Gouma is also the first person to receive National Science Foundation (NSF) I-Corps funding. She earned the support for a functionalized solar-powered "nanogrid" which breaks down hydrocarbons in polluted water (Cordova 2015). In addition to sensing applications, nanotechnology is enabling cheap, quick, and on-site DNA sequencing via nanopores. Lab-on-a-chip technology is bringing diagnostics to the field and, as microfluidics make way for nano-sized channels, increased sensitivity and detection will allow for even higher throughput screening of disease with less sample. Nanomaterials are also used in dental applications, tissue scaffolds for wound healing, antimicrobial bandages, imaging agents, and many other areas of medicine.

The work of Naomi Halas exemplifies the way nanotechnology spans multiple disciplines and applications (Halas 2018). She specializes in creating nanoparticles which have surface resonances across visible and infrared wavelengths. Her research in plasmonics and the ability to carefully control nanoparticles led her to develop the idea of the "tunable plasmon." The Halas lab has developed techniques to chemically manipulate the shape of nanoparticles which affect the collective electronic resonance, or plasmon, of the particle. Tuning the nanoscale geometry can change more than the color with applications in the broad area of metamaterials. Halas is driven by a desire that her work have "unique applications of societal and technological impact." This has led her to commercialize her responsive nanoparticles, starting companies specializing in cancer treatment and novel forms of energy generation. She is also part of a team, along with Qilin Li, commercializing technology with a light-harvesting membrane to heat and desalinate water in a single step (Rice University 2018; Li 2019).

In addition to desalination and collecting water from fog as noted in two of the examples above, the use of nanotechnology has the potential to help in other ways to improve access to clean, cheap water. Theresa Dankovich developed paper-based water filters infused with silver nanoparticles to allow for local water purification (Folia Water 2018). She started Folia Water with the goal of large-scale, low-cost production of the filters to enable people in developing countries to clean their own water. She also developed a method to produce the filters using nontoxic and renewable materials. In addition to providing energy for desalination and other methods of purification, nano-enabled sensors provide low-cost, low-energy methods to monitor water quality. The use of nanotechnology extends beyond the sustainable production and monitoring of clean water to the reclamation of polluted water. Scientists have developed nanoscale spongelike materials that soak up oil from water after a spill (Hashim et al. 2012; Barry et al. 2017). Additionally, nanocatalysts are being used to break down pollutants in water and soil.

Working to Ensure the Safe Development and Adoption of Nanotechnology

Along with research and development focused on applications, considerable attention has been devoted to the potential implications of nanotechnology. Two early voices in the area of ethics and societal perception were Rosalyn Berne and Barbara Herr Harthorn (Berne 2019; Harthorn 2019). Berne, who developed courses on nanoethics for engineers, used a 5-year CAREER grant from NSF to initiate a conversation on societal issues by interviewing 35 nanotechnology scientists and engineers. Her discussions are the basis for the book "NanoTalk" (Berne 2005). Harthorn, former Director of the NSF Center for Nanotechnology in Society, studied the perception of nanotechnological risk among experts and the general public in the USA and the UK. She also coedited the book "The Social Life of Nanotechnology" (Harthorn and Mohr 2012).

Responsible development of nanotechnology also includes the consideration of potential environmental, health, and safety (nanoEHS) implications. The body of knowledge gained in this area is immense and numerous women have contributed to the advancement of nanoEHS research including the development of tools and methods to evaluate nanomaterials and measure exposure. Several technical journals have been established which focus on these research areas, including NanoImpact where Socorro Vázquez-Campos serves as an associate editor (NanoImpact 2019). The US Government alone has invested over a billion dollars in this area of research; "Highlights of Recent Research on the Environmental, Health, and Safety Implications of Engineered Nanomaterials" includes more information (National Nanotechnology Initiative 2016). Worldwide collaborative efforts have been established, including the US-EU Communities of Research (US EU Nanotechnology Communities of Research 2018). In the EU, Eva Valsami-Jones and Iseult Lynch serve on the

Coordination team of the NanoSafety Cluster and recently published a reflection on the tools that have been developed for the risk assessment of nanomaterials (Fadeel et al. 2018; Lynch 2018; Valsami-Jones 2018). Jo Ann Shatkin has written extensively on the health and environmental risks of nanotechnology and recently identified key advancements in nanoEHS over the past 15 years (Shatkin 2012, 2018).

One important area that has advanced significantly is the safe handling of nano-materials in a laboratory or industrial setting. Resources for laboratory safety devel-oped by universities, government agencies, and organizations and professional societies have been assembled and promoted by the National Nanotechnology Coordination Office (National Nanotechnology Initiative 2019). The National Institute for Occupational Safety and Health (NIOSH) conducts research and field studies to help industry protect their workers (The National Institute for Occupational Safety and Health 2018). NIOSH Current Intelligence Bulletins that summarize state of the science and publications that provide approaches for safe handling of nanomaterials and workplace design suggestions are valuable resources for nano-technology researchers and developers alike.

Preparing Students, Workers, and the General Public

New nanotechnology discoveries and the development of innovative applications enabled by nanotechnology require scientists and engineers knowledgeable about the novel properties and behaviors at the nanoscale. Furthermore, public acceptance of nanotechnology is important for these new applications to be broadly adopted. To address these needs, educational programs for K-12, 2- and 4-year colleges, gradu-ate schools, and the general public have been developed in many regions of the world. The goal is to help build the workforce for future nanotechnology research and development, and to ensure an informed citizenry (National Nanotechnology Initiative n.d.; Malsch 2014; Asia Nano Forum 2018; Focus Nanotechnology Africa Inc. 2018). In an effort to benchmark efforts in nanoscale science and engineering education (NSEE), a workshop was held in 2010, followed by workshops in 2014 and 2017 (Murday 2010, 2014; Akbar et al. 2017). These workshops provided opportunities to assess current efforts, share best practices, and identify resources. One of the outcomes of these workshops was the assembly of hundreds of NSEE resources, from lesson plans and demonstrations to videos and hands-on experi-ments in a searchable database (Spadola and Friedersdorf 2017). A few of the many notable efforts in the area of NSEE are highlighted here.

The Nanoscale Informal Science Education Network (NISE Net) reached over 30 million people during its 10-year funding by the NSF (Bell 2015). This project, led by the Boston Museum of Science, established a national network of nearly 600 organi-zations including museums, universities, industry, and others. The NISE Network had three focus areas: educational deliverables (activities and programs, exhibits, media, and professional development), network infrastructure (project teams, regional hubs, and partners), and expanding the knowledge base of nanoscale science education.

The program established NanoDays, an annual event held at sites across the country, and the world, that highlighted nanoscale science and engineering with a focus on the current and future impacts of these areas on society (National Informal STEM Education Network 2018). A variety of demonstrations and other resources were developed during the active years of this program. Kits with instructions and materials for demonstrations were sent to educators at no cost to use during NanoDays. Although new kits are not being developed under this program, digital versions are available for download and many participants continue to use these resources during NanoDays and throughout the year in their educational and outreach efforts.

NanoDays helped to provide a focus on nanotechnology, but there are many other efforts in K-12 NSEE. Nancy Healy spent more than a decade teaching nanotechnology through outreach efforts into classrooms, summer camps, and, perhaps most importantly, K-12 teachers with Research Experience for Teachers programs, workshops, and boot camps (National Nanotechnology Coordinated Infrastructure 2018). Beyond reaching hundreds of teachers and thousands of students in her local community, Healy coordinated education and outreach efforts across the National Nanotechnology Coordinated Infrastructure, a nationwide network of nanotechnology user facilities based at 16 university sites across the country, and its predecessor the National Nanotechnology Infrastructure Network. The efforts of Julia Cothron, Mary Frances Hobbs, Daphne Schmidt, and Yvonne Pfluger at the MathScience Innovation Center in Richmond, Virginia, to develop and deliver nanotechnology to students and teachers were critical as the Commonwealth of Virginia was the first state in the USA to include nanotechnology in its K-12 science standards (MathScience Innovation Center 2018; Virginia Department of Education 2019).

Technicians are a critical piece of the nanotechnology workforce. Training occurs in a variety of ways including certificate programs and associate degrees through community or technical colleges. Deb Newberry established an Advanced Technology Center in Nanotechnology Education through an NSF grant called Nano-Link (Nano-Link Center for Nanotechnology Education 2018). Nano-Link has developed resources for educators and students and has worked closely with industry to assess their needs for trained technicians. Newberry also developed collaborations across the country to leverage resources and have a greater impact. The Nanotechnology Applications and Career Knowledge (NACK) Network, supported in part by NSF, also assists in technician training (Nanotechnology Applications and Career Knowledge Network 2019). One focus area of the NACK Network has been to provide students (and teachers) with access to cutting-edge equipment and cleanrooms through collaborations between colleges and universities. The Remotely Accessible Instruments for Nanotechnology (RAIN) partnership enables remote access to equipment like electron microscopes where students can control the instrument from their classroom (NACK Network 2019).

At the university level, nanotechnology classes are now commonplace in engineering and science departments and many certificate, minor, and degree programs have been established (National Nanotechnology Initiative n.d.). Students interested in any area of science, technology, engineering, and mathematics (STEM) will likely encounter nanotechnology. To encourage K-12 students on this path, the novel properties at the nanoscale can be used to inspire them to pursue a STEM education. One way to excite students is to tie current and future applications to superpowers. For example, the nanoscale structure of gecko feet that enables them to scamper up walls has been inspirational for students and developers alike. Along with metamaterials enabling invisibility cloaks, bulletproof CNT textiles, and ultra-strong lightweight cables for swinging, gecko-inspired gloves can round out the gear for any burgeoning superhero!

Summary

Nanotechnology is already impacting nearly every aspect of daily life. Stain-resistant pants; the powerful computer in mobile devices; touch screens; water-resistant electronics; brighter yet more energy-efficient displays; sunscreens and cosmetics; sports equipment; paints and protective coatings; antimicrobial bandages; lightweight yet stronger materials to protect warfighters; electromagnetic shielding for electronics in space; more efficient food packaging; and sensors with uses in medicine, food, and air quality are just a few examples of applications enabled through nanotechnology.

The women highlighted in this chapter and throughout this book are just a few of the many who have made significant contributions across a wide variety of science and engineering fields through their work in nanotechnology. As progress and understanding grow, improvements to existing products will give way to whole new applications and methods enabled by nanotechnology. The talented girls and young women that follow in this exciting field will bring this promise to fruition.

References

Advanced Silicon Group. (2018). *Meet our team*. Retrieved December 2018, from http://www.advancedsilicongroup.com/our-team.html.

Akbar, B., Brummet, J. L., Flores, S. C., Gordon, A., Gray, B., & Murday, J. S. (2017). *Global perspectives in convergence education* (Workshop report). Washington, DC.

American Physical Society. (2018). Mildred "Millie" Dresselhaus. *Explore the Science*. Retrieved December 29, 2018, from http://www.physicscentral.com/explore/people/dresselhaus.cfm.

Asia Nano Forum. (2018). *Education*. Retrieved January 3, 2019, from https://www.asia-anf.org/working-groups/education/.

Barry, E., Mane, A. U., Libera, J. A., Elam, J. W., & Darling, S. B. (2017). Advanced oil sorbents using sequential infiltration synthesis. *Journal of Materials Chemistry A, 5*(6), 2929–2935.

Belcher, A. (2012). *The biomolecular materials groups at MIT*. Retrieved December 29, 2018, from http://belcherlab.mit.edu/.

Bell, L. (2015) *Nanoscale informal science education network legacy and future plans*. Retrieved May 13, 2019 from http://www.nseresearch.org/2015/presentations/Larry_Bell~30_Bell_NSE_Grantee_2015_reduced.pdf.

Berne, R. W. (2005). *Nanotalk*. Boca Raton, FL: CRC Press.

Berne, R. W. (2019). Retrieved January 3, 2019, from https://engineering.virginia.edu/faculty/rosalyn-w-berne.

Bradbury, M. (2018). *The Michelle Bradbury lab*. Retrieved December 21, 2018, from https://www.mskcc.org/research-areas/labs/michelle-bradbury.

Cordova, F. (2015). An overview of the budget proposals for the National Science Foundation and National Institute of Standards and Technology for fiscal year 2016. In *Subcommittee on research and technology (114th Congress)*. Committee on Science, Space, & Technology.

Fadeel, B., Farcal, L., Hardy, B., Vázquez-Campos, S., Hristozov, D., Marcomini, A., Lynch, I., Valsami-Jones, E., Alenius, H., & Savolainen, K. (2018). Advanced tools for the safety assessment of nanomaterials. *Nature Nanotechnology, 13*(7), 537–543.

Focus Nanotechnology Africa Inc. (2018). *FONAI home*. Retrieved January 3, 2019, from http://fonai.org/Home_Page.php.

Folia Water. (2018). *Scientific innovation*. Retrieved December 21, 2018, from https://www.foliawater.com/silver-papers/.

Glotzer, S. (2018). *The Glotzer group*. Retrieved December 29, 2018, from http://glotzerlab.engin.umich.edu/home/.

Gouma, P.-I. P. (2018). *About the director*. Retrieved December 21, 2018, from https://acrl.osu.edu/people/about-director.

Greer, J. R. (2018). *Greer group*. Retrieved December 21, 2018, from http://www.jrgreer.caltech.edu/people/jrgreer.html.

Halas, N. (2018). *Halas research group*. Retrieved December 2018, from http://halas.rice.edu/.

Hammond, P. T. (2015). *The Hammond lab*. Retrieved December 29, 2018, from https://hammondlab.mit.edu/.

Harthorn, B. H. (2019). Retrieved January 3, 2019, from http://www.anth.ucsb.edu/people/barbara-herr-harthorn.

Harthorn, B. H., & Mohr, J. W. (2012). *The social life of nanotechnology*. New York: Routledge.

Hashim, D. P., Narayanan, N. T., Romo-Herrera, J. M., Cullen, D. A., Hahm, M. G., Lezzi, P., Suttle, J. R., Kelkhoff, D., Muñoz-Sandoval, E., Ganguli, S., Roy, A. K., Smith, D. J., Vajtai, R., Sumpter, B. G., Meunier, V., Terrones, H., Terrones, M., & Ajayan, P. M. (2012). Covalently bonded three-dimensional carbon nanotube solids via boron induced nanojunctions. *Scientific Reports, 2*.

Iijima, S. (1991). Helical microtubules of graphitic carbon. *Nature, 354*(6348), 56–58.

Kreibig, U., & Vollmer, M. (1995). *Optical properties of metal clusters*. Berlin Heidelberg: Springer.

Kroto, H. W., Heath, J. R., O'Brien, S. C., Curl, R. F., & Smalley, R. E. (1985). C_{60}: Buckminsterfullerene. *Nature, 318*(6042), 162–163.

Li, Q. (2019). *Qilin's research group*. Retrieved January 16, 2019, from https://ceve.rice.edu/qilin-li.

Lynch, I. (2018). Retrieved January 2019, from https://www.birmingham.ac.uk/schools/gees/people/profile.aspx?ReferenceId=56477.

Malsch, I. (2014). Nano-education from a European perspective: Nano-training for non-R&D jobs. *Nanotechnology Reviews, 3*(2), 211–221.

MathScience Innovation Center. (2018). *About us*. Retrieved January 8, 2019, from https://www.mymsic.org/about-us.

Meza, L. R., Das, S., & Greer, J. R. (2014). Strong, lightweight, and recoverable three-dimensional ceramic nanolattices. *Science, 345*(6202), 1322–1326.

Murday, J. (2010). *International benchmark workshop on K-12 nanoscale science and engineering education*.

Murday, J. (2014). *Nanoscale Science and Engineering Education (NSEE)—The next steps workshop report*.

NACK Network. (2019). *Remote access*. Retrieved from http://nano4me.org/remoteaccess.

Nam, K. T., Kim, D. W., Yoo, P. J., Chiang, C. Y., Meethong, N., Hammond, P. T., Chiang, Y. M., & Belcher, A. M. (2006). Virus-enabled synthesis and assembly of nanowires for lithium ion battery electrodes. *Science, 312*(5775), 885–888.

NanoImpact. (2019). *Socorro Vázquez-Campos*. Retrieved January 3, 2019, from https://www.journals.elsevier.com/nanoimpact/editorial-board/dr-socorro-vazquez-campos.

Nano-Link Center for Nanotechnology Education. (2018). Retrieved January 3, 2019, from https://www.nano-link.org/.

Nanotechnology Applications and Career Knowledge Network. (2019). Retrieved from http:// nano4me.org/.

National Informal STEM Education Network. (2018). *NanoDays*. Retrieved from http://www. nisenet.org/nanodays.

National Nanotechnology Coordinated Infrastructure. (2018). *Nancy Healy*. Retrieved January 4, 2019, from https://www.nnci.net/experts/nancy-healy.

National Nanotechnology Initiative. (2016). *Highlights of recent research on the environmental, health, and safety implications of engineered nanomaterials*. Retrieved January 2019, from https://www.nano.gov/sites/default/files/pub_resource/Highlights_Federal_NanoEHS_ FINAL.pdf.

National Nanotechnology Initiative. (2019). *Resources for nanotechnology laboratory safety*. Retrieved January 3, 2019, from https://www.nano.gov/LabSafety.

National Nanotechnology Initiative. (n.d.). *Education*. Retrieved January 2, 2019, from https:// www.nano.gov/education-training.

NBD Nano. (2018). Retrieved December 21, 2018, from https://www.nbdnano.com/.

Novoselov, K. S., Geim, A. K., Morozov, S. V., Jiang, D., Zhang, Y., Dubonos, S. V., Grigorieva, I. V., & Firosov, A. A. (2004). Electric field effect in atomically thin carbon films. *Science, 306*(5696), 666–669.

Rice University. (2018). *Freshwater from salt water using only solar energy*. Retrieved May 13, 2019 from http://news.rice.edu/2017/06/19/freshwater-from-salt-water-using-only-solar-energy-2/.

Royal Swedish Academy of Sciences. (2010). *The Nobel Prize in physics*. Retrieved May 13, 2019 from https://www.kva.se/en/pressrum/pressmeddelanden/nobelpriset-i-fysik-2010.

Shatkin, J. A. (2012). *Nanotechnology: Health and environmental risks*. Boca Raton, FL: CRC Press.

Shatkin, J. A. (2018). *15 (or so) things we've learned in 15 years of the national nanotechnology initiative—the view from Vireo Advisors*. Retrieved May 13, 2019 from http://www.vireoadvisors.com/blog/2018/12/4/15-or-so-things-weve-learned-in-15-years-of-the-national-nanotechnologyinitiative-the-view-from-vireo-advisors.

Spadola, Q., & Friedersdorf, L. (2017). *Nano education resources*. Retrieved May 13, 2019 from https://nanohub.org/publications/118/1.

The National Institute for Occupational Safety and Health. (2018). *Nanotechnology*. Retrieved January 3, 2019, from https://www.cdc.gov/niosh/topics/nanotech/.

US EU Nanotechnology Communities of Research. (2018). Retrieved January 3, 2019, from https://us-eu.org/communities-of-research/overview/.

Valsami-Jones, E. (2018). Retrieved January 2019, from https://www.birmingham.ac.uk/staff/profiles/gees/valsami-jones-eva.aspx.

Virginia Department of Education. (2019). *Standards of learning documents for science—Adopted 2010*. Retrieved January 3, 2019, from http://www.doe.virginia.gov/testing/sol/standards_ docs/science/index.shtml#sol.

Woggon, U. (1997). *Optical properties of semiconductor quantum dots*. Berlin Heidelberg: Springer.

Yu, M. F., Lourie, O., Dyer, M. J., Moloni, K., Kelly, T. F., & Ruoff, R. S. (2000). Strength and breaking mechanism of multiwalled carbon nanotubes under tensile load. *Science, 287*(5453), 637–640.

Chapter 2
Nanotechnology as a Tool for Science and Scientific Literacy

Saniya LeBlanc

Introduction

Nanotechnology education efforts have resulted in resources for formal, informal, and nonformal education of a wide a variety of audience. There are now many mechanisms for spreading knowledge about nanotechnology. Examples include grades K-12 workshops, undergraduate courses, textbooks, museum exhibits, web sites, and much more (Roco 2002, 2003a; Greenberg 2009; Bach and Waitz 2015). There have been many studies about the effectiveness of different educational approaches and tools in increasing nanotechnology awareness, knowledge, and skills. The overarching goal has been nanotechnology literacy—achieved through science education tools. This chapter discusses an alternative perspective about nanotechnology education. It explores the use of nanotechnology as a literacy tool; in this inversion, nanotechnology becomes an effective tool for achieving science/ scientific literacy without concern for specific nanotechnology concepts. Given the broader societal goals for science/scientific literacy, there is a need for effective tools to achieve that literacy. Perhaps nanotechnology is an ideal topic to achieve these improvements.

Laherto's investigation of nanotechnology's educational significance explored this approach. Laherto discussed how nanotechnology can enhance science education by providing modern conceptions about science (Laherto 2010). Using a Model of Educational Reconstruction, Laherto explored whether it is worthwhile to teach a particular science topic (i.e., nanoscience) and whether it is possible to do so (Laherto 2010). However, much like the extensive body of nanotechnology education research, the science education perspective alone cannot predict nanotechnology's potential as an educational tool. Science education (how people

S. LeBlanc (✉)
Department of Mechanical & Aerospace Engineering, The George Washington University, Washington, DC, USA
e-mail: sleblanc@gwu.edu

© Springer Nature Switzerland AG 2020
P. M. Norris, L. E. Friedersdorf (eds.), *Women in Nanotechnology*, Women in Engineering and Science, https://doi.org/10.1007/978-3-030-19951-7_2

Fig. 2.1 The diagram shows the four categories and their intersections discussed in this chapter about nanotechnology as an educational tool for science and scientific literacy of STEM audiences and the general public

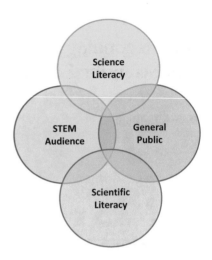

learn and what they should learn) must be combined with an understanding of perceptions and attitudes.

To investigate nanotechnology as a tool for science/scientific literacy, it is first necessary to understand the intersection of nanotechnology, science/engineering education, and communications. As depicted in Fig. 2.1, there are two literacy goals and two audiences in which this intersection is considered here. *Science* literacy (science knowledge) is differentiated from *scientific* literacy (critical thinking developed from scientific understanding), and two audience are distinguished based on their technical needs. The technical audience consists of people who regularly engage with science, technology, engineering, and math (STEM) topics. Members of the general public do not have extensive STEM training nor do they engage regularly with STEM topics. These four categories are described in more detail in later sections.

The discussion presented here does not parse the differences between formal, informal, and nonformal education. To the extent these terms are used here, they refer to the level of structure and teacher-led activity and instruction. Formal education is used to denote structured learning environments with instruction led by a teacher or guide. Informal education refers to unstructured environments in which the learner is self-guided.

The discussion presented here treats nanoscale science and engineering collectively and uses the term "nanotechnology" for simplicity. While there are certainly differences between nanoscience and nanotechnology, parsing the differences would detract from the broader investigation around science/scientific literacy. Moreover, the collective consideration of both science and technology is important to this discussion since scientific literacy could not be achieved by technology alone (Maienschein and Students 1999); measures to improve literacy must consider technology and science as well as the link between the two.

At its core, this chapter explores the question: Can nanotechnology education improve the science and scientific literacy of technical and nontechnical audiences?

The question provides a lens through which to view and interpret existing research findings at the intersection of pertinent fields: science/engineering, education, social science, and communications. The discussion indicates nanotechnology is an effective educational tool to improve both science and scientific literacy for technical (STEM) audiences. By contrast, nanotechnology education is probably not an effective tool for the general public's *scientific* literacy, but nanotechnology knowledge is necessary for contemporary *science* literacy. Regardless of the audience, framing and the narrative created around nanotechnology are pivotal to public perceptions and attitudes, even at the expense of the content knowledge itself.

Relevance for Nanotechnology Practitioners

The target reader for this chapter consists of scientists and engineers in the field of nanotechnology as well as science educators. Often, these are one and the same (e.g., professors engaged in research and teaching). The relevance for educators is perhaps the clearest. Educators are tasked with improving science/scientific literacy, so new, effective educational tools are useful. There are a myriad of topics (e.g., robotics, biotechnology, additive manufacturing) which an educator could choose to engage an audience. If nanotechnology can be a particularly effective educational tool for improving science/scientific literacy, then it is worthwhile for educators to consider adding it to their repertoire.

Nanotechnology scientists and engineers regularly communicate their work to others. Some may primarily communicate nanotechnology to technical peers or scientists/engineers in training, so this paper's discussion of nanotechnology to improve science literacy of STEM audiences will be useful. However, many scientists and engineers are also concerned with improving the science/scientific literacy of those beyond the technical sphere. They are motivated by a belief that knowledge—even basic knowledge—about science and scientific processes is necessary for the public. Moreover, the critical thinking associated with scientific literacy is necessary to participate in society. For these scientists and engineers, perhaps nanotechnology education is a useful tool to achieve these goals. Nanotechnology experts are uniquely poised to choose what knowledge they communicate. Because nanotechnology is highly interdisciplinary (Porter and Youtie 2009), experts have disciplinary expertise in specific topics as well as interdisciplinary experience. They can choose to communicate disciplinary content (e.g., chemistry, electrical engineering) or the interdisciplinary perspectives resulting from nanotechnology. Perhaps that choice should be made to improve scientific literacy (rather than impress the audience, for example).

The discussion presented here draws on science, engineering, education, social science, and communications research to explore whether nanotechnology education could be a tool to improve science and scientific literacy. The motivation is to link practitioners—educators, scientists, and engineers who are actively creating and communicating technical content—with science communication and nanotechnology education research findings. As such, this discussion will surely offend by failing to use terminology and theories specific to education, social science, and communication research. To some extent, it is intentional since the chapter discusses the topic in a manner accessible to practitioners who are on the front lines of improving science and scientific literacy both formally and informally.

Science Versus Scientific Literacy

The difference between science literacy and scientific literacy is critical to this discussion. Technological literacy is often perceived and defined differently than science/scientific literacy; however, it is lumped with science/scientific literacy here because science/scientific literacy has been clearly articulated and can encompass technological literacy for the goals of this discussion. Comparisons of science and scientific literacy have been made elsewhere (Maienschein and Students 1999; Laherto 2010), and the focus here is on their relationship to nanotechnology education. Science literacy refers to content knowledge and technical skills. Improving science literacy is often motivated by a need to create and train a technical workforce and a belief that all members of society should have a basic knowledge of core science topics and concepts. In other words, everyone should have basic science literacy, and those in the STEM workforce should have more advanced science literacy.

To date, nanotechnology education has largely been framed within the context of science literacy. As nanotechnology emerged and grew, there was a need for a workforce trained to participate in the field (Roco 2002). The resulting nanotechnology courses, training programs, and degrees or certifications are all examples of formal education designed to inspire, educate, and train the workforce needed for research, development, commercialization, and regulation of nanotechnology. The economic, environmental, and safety impacts of nanotechnology led to a need for the broader public to have a basic awareness and understanding of the emerging technologies. Informal education through media exposure, museum exhibits, and outreach events aims to increase the nanotechnology science literacy of the broader public.

By contrast, scientific literacy has a higher cognitive aim: understanding scientific principles and processes and then applying this understanding to relevant societal issues. Scientific comprehension and analysis thus lead to critical thinking skills for technical topics. A participatory democracy in which citizens contribute to the overall progress of society requires those citizens to have a high level of scientific literacy. Even if citizens do not have content knowledge on certain topics, they should have the analysis, evaluation, and synthesis skills to interpret technical content and make assessments and decisions about how it impacts society. The American Association for the Advancement of Science makes the case for both science and scientific literacy, and its hallmark publication *Science for All Americans* articulates a clear argument for scientific literacy:

Scientific habits of mind can help people in every walk of life to deal sensibly with problems that often involve evidence, quantitative considerations, logical arguments, and uncertainty; without the ability to think critically and independently, citizens are easy prey to dogmatists, flimflam artists, and purveyors of simple solutions to complex problems (American Association for the Advancement of Science 1990).

Acknowledging the difficulty with defining scientific literacy, Maienschien and her students presented five principles underlying the attributes of U.S. scientific literacy: scientific literacy is "(1) a desirable goal, (2) for all Americans, (3) measurable and assessable, (4) useful for everyday life, and (5) inextricably connected with its social context" (Maienschein and Students 1999). She discussed it as an "intrinsic good" as opposed to the "instrumental good" of science literacy (Maienschein and Students 1999).

There have been far fewer studies explicitly investigating the scientific literacy impact of nanotechnology education. Activities which aim to connect nanotechnology to societal issues ultimately impact scientific literacy. However, much of the nanotechnology education aimed at science literacy (increasing content knowledge about nanotechnology) may in fact improve scientific literacy, even though this impact is not evaluated. Indeed, many nanotechnology educators have likely witnessed this connection. The concepts of bottom-up and top-down fabrication are a good example. To witness an engineering student grapple with the concepts of self-assembly or nano/microfabrication using photolithography, deposition, and etch processes is to witness a complete rewiring of that person's understanding of manufacturing, connections between fundamental science and engineering processes, and everyday products and devices. The resulting analytical skills could be applied to other topics/contexts and constitute an improvement in scientific literacy. Future research could evaluate *scientific* literacy changes resulting from nanotechnology education activities which aim to increase *science* literacy.

Target Audience

Nanotechnology education can improve both science and scientific literacy, but it is still not clear whether nanotechnology is an effective tool for improving science and scientific literacy when nanotechnology education is not the end goal. Target audience is likely a critical factor. Target audience is divided into two broad categories: technical and general. The technical audience consists of the current and future STEM workforce. It includes people using STEM knowledge and skills for their current endeavors. Members of this audience are well-versed in science and technology, and they engage with technical content on a regular basis. Much of the nanotechnology education activities have focused on this audience (Roco 2002, 2003a; Greenberg 2009). The general audience ("the public") is everyone not in the technical audience. The general public should be educated about nanotechnology since it impacts many aspects of society (Roco 2003b; Laherto 2010; National Nanotechnology Initiative 2016). Nanotechnology education research focusing on the general public has investigated what mechanisms are most effective for increasing nanotechnology literacy (National Informal STEM Education Network n.d.). Nanotechnology education can only be achieved with nanotechnology content, but there are other ways to achieve science/scientific literacy than nanotechnology. Does nanotechnology offer particular advantages for achieving literacy for either audience? The following discussion evaluates both audiences.

Technical (STEM) Audience

Improving the science/scientific literacy of the technical audience is the most straightforward consideration. As discussed previously, nanotechnology education can improve the science literacy of this audience, particularly in secondary and higher education settings. Furthermore, recruiting and training a STEM workforce with nanotechnology knowledge and skills are imperative, so there have been extensive efforts to meet these education needs with nanotechnology education (Roco 2002, 2003a; Malsch 2008; Bach and Waitz 2015; National Nanotechnology Initiative 2016).

In addition to the STEM workforce needs, nanotechnology is intimately linked to societal issues (Roco 2003b), so it is an opportune topic to improve scientific literacy. There are multiple examples demonstrating the efficacy of nanotechnology as a tool for improving the scientific literacy of science and engineering students. Nanotechnology courses have been used to teach the ethical, legal, monetary, and social issues associated with emerging technologies (Porter 2007; Hoover et al. 2009; Wang et al. 2013; LeBlanc et al. 2016a, b). The courses have been valuable forums in which to teach critical thinking. For example, two bio-nanotechnology courses had students analyze technical literature in order to interpret results, detect inconsistencies, verify alignment with physical laws, and recognize exaggerations of results and impacts (Wang et al. 2013). One introductory nanotechnology course explicitly identified and dispelled misconceptions about nanotechnology (Porter 2007). Notably, this undergraduate course was open to students from any major, so it provides a good example of using nanotechnology to improve scientific literacy of both technical and nontechnical audiences. It also demonstrates the extensive effort required to improve scientific literacy. Students had to identify prior knowledge and mindsets, characterize how topics are framed by media, and develop critical thinking skills to link their own understanding of the science to the social context in which the science would be implemented. Another course explicitly linked nanotechnology's social and ethical issues to the fundamental science and technological developments, resulting in a majority of students believing nanotechnology researchers should receive ethics training (Hoover et al. 2009). While nanotechnology can improve scientific literacy, and it is effective at doing so, it requires extensive effort to construct the necessary knowledge and rationally connect it to the social context.

The interdisciplinary nature of nanotechnology (Porter and Youtie 2009) makes it a particularly effective tool for improving scientific literacy. Nanotechnology topics can be used to demonstrate how knowledge from multiple fields combines to create new understanding and capabilities. Nanotechnology also demonstrates how activities in science and technology impact societal and ethical issues (Roco 2003b). For instance, manufacturing nanomaterials can introduce nanoparticulates into the environment, causing health and safety concerns. Nanotechnology's applicability to other disciplines can also provide an entry point for people to understand the connection between their own discipline and other fields. Notable examples use art such as stained glass (Duncan et al. 2010) and virtual reality (Schönborn et al. 2014) to engage audiences. Nanotechnology's interdisciplinary nature demonstrates its broad relevance and applicability, and its multiple dimensions can demonstrate to students the many pathways available to them in science and engineering. This may improve recruitment and retention in STEM fields.

The discussion thus far has been relevant for STEM students, since most of the mechanisms described require a curricular component. This leaves the adult technical audience in a situation similar to the general public. While the adult technical audience has more technical prior knowledge, its opportunities to be exposed to nanotechnology are similar to those of the general public. However, once exposed, its technical training will likely make scientific literacy improvement efforts more effective.

General Public

The general population should have basic nanotechnology knowledge, so the discussion of whether nanotechnology can improve science literacy becomes circular because contemporary science literacy must include nanotechnology (Laherto 2010). Nanotechnology education necessarily improves science literacy since it teaches and improves understanding of standard or core science and broadens this understanding to include nanotechnology. However, a nanotechnology education activity's capacity to improve science literacy may depend on the education mechanism (formal or informal) and the segment of the general public being targeted.

The "general public" audience for nanotechnology education can be considered in two groups: grades K-12 students and adults. Both formal and informal education activities have been implemented for K-12 audiences while nanotechnology education for adult audiences has been largely through informal education. For the K-12 audience, nanotechnology education research results demonstrate the most effective mechanisms for improving nanotechnology literacy. The NanoLeap project by the Mid-Continent Research for Education and Learning explicitly linked core concepts, transitional ideas, and nanotechnology concepts in grades 7–12 (*McREL: Education and Public Outreach: NanoLeap: Home Page* n.d.; Greenberg 2009).

Interestingly, the NanoSense project led by SRI International demonstrated how nanotechnology presented without the explicit connection to the core science concepts was ineffective at achieving meaningful understanding (Greenberg 2009). Nanotechnology education research results not only indicate nanotechnology education *can* improve science literacy, but they also show *how* to do so.

The intersection between science education and communications is critical for influencing the general public, particularly adult audiences whose literacy is impacted by informal education mechanisms (e.g., media outlets, museum exhibits) (National Informal STEM Education Network n.d.). One study found most people were exposed to nanotechnology through mass media outlets, and there was a significant gap between what technical experts assume people know and the public's actual level of prior knowledge (Castellini et al. 2007). The communication medium affects both nanotechnology knowledge and public attitudes (Lee and Scheufele 2006). One study found a positive link between television, newspaper, and web science media and public attitudes, but people's nanotechnology knowledge depends on the communication medium. Newspaper and web science use were positively linked to nanotechnology knowledge. However, television science use had a positive link to "deference toward scientific authority;" people who rely on gathering science knowledge from television form their attitudes based on what the scientific authority presents (Lee and Scheufele 2006).

Communications research yields another informative result: content knowledge is less relevant than the nonscientific frameworks people use to interpret new knowledge. New literacy develops within the context of people's social narratives or their prior mindsets, and these mindsets shape the new knowledge (Feinstein 2015). This shaping of perspectives has been demonstrated specifically for nanotechnology by investigating the interaction of cognitive and affective influences on public attitudes towards nanotechnology (Lee et al. 2005). Cognitive influence is the impact of science knowledge, and affective influence is the effect of feelings like concern or fear. A 2004 study investigated how knowledge and emotion impact support for nanotechnology and perceived risks and benefits: people—both experts and laypeople—make judgments based on their feelings about science and technology rather than analytical assessments (Lee et al. 2005). Even increased nanotechnology literacy did not lead to more public support for nanotechnology (although this finding contradicts existing literature about science knowledge and support for scientific issues) (Scheufele and Lewenstein 2005). The study used a "cognitive miser" model to interpret the influence of media on public attitudes towards nanotechnology. Using cognitive shortcuts, "people only collect as much or as little information about a given issue as they think is necessary to make a decision" (Scheufele and Lewenstein 2005). In taking these shortcuts, people interpret news based on how media or others frame it. Increased knowledge about nanotechnology (i.e., better science literacy) did not improve public attitudes towards the technology; people relied on other sources to frame their understanding (Scheufele and Lewenstein 2005). The complexity of nanotechnology may steer people more strongly towards using these cognitive shortcuts because it would require considerable effort to gain deep understanding about the topic.

Since scientific literacy is characterized by understanding the use of science in everyday life and within social contexts, these studies yield insights into using nanotechnology for scientific literacy improvements. The foremost factor is acknowledging and creating the social narrative and mindset: the framing of nanotechnology will directly influence the resulting scientific literacy. The complexity of the topic will prove challenging because of the public's tendency to make decisions with as little information as possible (Scheufele and Lewenstein 2005). The informal education outlets used by the public are characterized by short timescales, relatively short, discrete time segments in which people are encountering the topic. Negotiating these time constraints with nanotechnology's complex content will be a crucial consideration.

The Role of Technical Experts

Perhaps the strongest understanding about scientific literacy comes from considering the interactions between the technical audience and the general public. Feinstein used discussions about the nature of "the public" to consider this interaction (Feinstein 2015). The works of Walter Lippmann and John Dewey describe the public as a political entity composed of citizens who participate in a discourse about the issues relevant to their society. In this conception, most citizens do not—and need not—have specialized knowledge for specific societal issues. The very existence of specialization means individuals do not have knowledge from a specialization other than their own. People have many components of their everyday lives with which to contend, so they will not contend with specialized details beyond their own expertise (Feinstein 2015).

Science practitioners and educators may find this perspective illuminating (albeit painful to accept) since it means science literacy need not be a goal. The public is not inclined to acquire new science knowledge, and its acquisition of that knowledge is unnecessary. Nonetheless, the perspective does not dispense with scientific literacy, and the interaction between experts and the public becomes a differentiating feature. In Fig. 2.2, Feinstein synthesized two different views. In Lippman's approach, the public elects decision-makers, and experts advise decision-makers. In Dewey's approach, there is a back-and-forth dialogue between experts and the public, both of whom influence policy (Feinstein 2015).

The Lippman approach places enormous responsibility on elected decision-makers, so these decision-makers must have strong scientific literacy. In this case, nanotechnology education could be an effective tool for improving scientific literacy. Its interdisciplinary nature, reliance on linking fundamental science to emerging technologies, and breadth of societal impact make nanotechnology education an ideal medium to improve the scientific literacy of decision-makers. However, the complexity of the topic requires these decision-makers do not take "cognitive shortcuts" (Scheufele and Lewenstein 2005) nor rely on "affective heu-

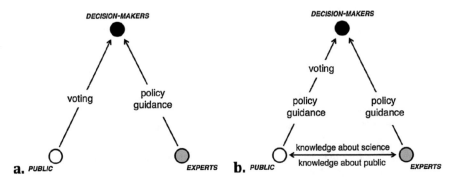

Fig. 2.2 Diagrams of the relationship between experts, the public, and decision-makers as recommended by (**a**) Lippman (left) and (**b**) Dewey (right). Reproduced with permission from Feinstein (2015)

ristics" (Lee et al. 2005) to make decisions. They would need to participate in formal education from experts.

With the Dewey approach, both the public and decision-makers need strong scientific literacy. Mutual exchange of knowledge between experts and the public must occur through formation of communities in which there is dialogue and debate (Feinstein 2015). Individual scientific literacy is unimportant compared to the collective scientific literacy of the community. It is challenging to use nanotechnology as a topic with which to improve this collective, community-based scientific literacy. The field's complexity and many facets could prove overwhelming in settings where informal education is combined with civic debate. On the other hand, the multifaceted complexity could be the key to its potency as an educational tool. Nanotechnology's interdisciplinary features and broad impact demonstrate vested interests of different stakeholders. Dialogue between these stakeholders and the general public could improve the public's critical analysis ability even if the public does not grasp the detailed knowledge behind each stakeholder's interests. Thus, nanotechnology education and communications research provide insight into how to engage in this dialogue. In particular, the framing used by experts in this dialogue will be critical to its impact on scientific literacy. Future research and analysis could investigate methods for effective framing.

Regardless of which perspective accurately describes the relationship between the public, decision-makers, and experts, the critical role of experts—in this case, nanotechnology experts—is clear. Nanotechnology scientists and engineers are gatekeepers for scientific literacy, and thus they hold the same responsibility as science educators. The importance of the experts' voice arises repeatedly in research on public knowledge, perceptions, and attitudes. In one study, trust in scientists was positively linked to general support for nanotechnology; in fact, someone's trust in scientists was a better predictor of support for nanotechnology than his/her own science knowledge (Lee et al. 2005).

Takeaways for Educators, Scientists, and Engineers

Educators, scientists, and engineers can gain insight and direction from analyzing the intersection of their fields with research results about communications and media. Some key insights emerge related to nanotechnology education. For all audiences, nanotechnology education is effective for improving *science* literacy since contemporary science knowledge should include knowledge about nanotechnology. Nanotechnology education seems most effective at improving *scientific* literacy when implemented in formal learning environments. Educators can use nanotechnology as a powerful educational tool for high school and college students because it integrates a wide array of technical disciplines and has social, ethical, and financial dimensions. In informal education environments and interactions with the general public, the framing of scientific issues is crucial since people will make decisions with minimal specialized knowledge. Communicating nanotechnology knowledge may be a desirable goal, but it is more effective to influence the narrative about science/engineering and emerging technologies as well as their societal contexts. A disconcerting conclusion emerges: perhaps improving the public's scientific literacy must be done at the expense of its science literacy in order to achieve the "intrinsic good" of a judicious populace.

How should educators and nanotechnology experts aim for scientific literacy in their communications? One approach may be to select which concepts to communicate based on the audience. Table 2.1. summarizes the "Big Ideas in Nanoscale Science and Engineering" along with the audience for which they were originally outlined (Greenberg 2009; Stevens et al. 2009). The ideas prescribed for grades 7–12 are presumably appropriate for the general public, and the ideas specified for grades 13–16 can apply to technical audiences. Given the challenges with communicating the breadth of ideas, which ideas, at a minimum, need to be communicated to improve literacy? In an effort to spark discussion and debate, the table presents one possibility. For the general public, perhaps the pertinent concept is size and scale combined with an understanding that emerging technologies dealing with the nanoscale have certain societal impacts. The overarching vision is to unite science, engineering, education, and communications expertise to design media which improve scientific literacy—whether the topics selected in rightmost columns of Table 2.1. can do so is moot. Future discussions could expand this table to include "big ideas" from other fields. The resulting comparison between nanotechnology and other emerging technologies would indicate which topics are most appropriate for improving scientific literacy.

Nanotechnology researchers have an important role to play in improving scientific literacy. As technical experts, the public relies on them to engage in a dialogue, but they must carefully consider how and to whom they are presenting information. What and how they present the information—and whether nanotechnology is the right medium—are relevant considerations. If scientific literacy is imperative, then scientists, engineers, and educators should engage communications experts in the quest to improve scientific literacy. Maybe nanotechnology is one tool they can all use in this quest.

Table 2.1. Suggested minimum concepts required for science and scientific literacy of the general public and STEM audiences presented within the framework of "Big Ideas in Nanoscale Science and Engineering" education (Stevens et al. 2009)

Audience	Big ideas of nanotechnology	Minimum required for	
		Science literacy	Scientific literacy
Grades 7–12 and general public	Size and scale	✓	✓
	Properties of matter	✓	
	Particulate nature of matter	✓	
	Modeling		
	Dominant forces	✓	
	Tools		
	Self-assembly		
	Technology and society		✓
Grades 13–16 and STEM audience	Size and scale	✓	✓
	Size-dependent properties	✓	✓
	Tools and instruments/ characterization	✓	
	Models and simulations		
	Surface-dominated behavior		
	Self-assembly		
	Surface-to-volume ratio	✓	✓
	Quantum mechanics	✓	✓
	Societal impact/public education		✓

Acknowledgements The author would like to acknowledge support from the National Science Foundation (EEC-1446001, NUE: An Interdisciplinary Practicum Approach to Nanotechnology Curricula Integration). The views expressed here are those of the author and do not necessarily reflect the views of the National Science Foundation.

References

American Association for the Advancement of Science. (1990). *Science for all Americans*. New York: Oxford University Press. Retrieved September 30, 2017, from http://www.project2061.org/publications/sfaa/online/intro.htm.

Bach, A.-M., & Waitz, T. (2015). International activities in nanoscale science and engineering education. In *New perspectives in science education*. Retrieved September 29, 2017, from https://conference.pixel-online.net/npse2013/common/download/Paper_pdf/034-STM03-FP-Bach-NPSE2013.pdf.

Castellini, O. M., et al. (2007). Nanotechnology and the public: Effectively communicating nanoscale science and engineering concepts. *Journal of Nanoparticle Research, 9*(2), 183–189. https://doi.org/10.1007/s11051-006-9160-z.

Duncan, K. A., et al. (2010). Art as an avenue to science literacy: Teaching nanotechnology through stained glass. *Journal of Chemical Education, 87*(10), 1031–1038. https://doi.org/10.1021/ed1000922.

Feinstein, N. W. (2015). Education, communication, and science in the public sphere. *Journal of Research in Science Teaching, 52*(2), 145–163. https://doi.org/10.1002/tea.21192.

Greenberg, A. (2009). Integrating nanoscience into the classroom: Perspectives on nanoscience education projects. *ACS Nano, 3*(4), 762–769. https://doi.org/10.1021/nn900335r.

Hoover, E., et al. (2009). Teaching small and thinking large: Effects of including social and ethical implications in an interdisciplinary nanotechnology course. *Journal of Nano Education, 1*(1), 86–95. https://doi.org/10.1166/jne.2009.013.

Laherto, A. (2010). An analysis of the educational significance of nanoscience and nanotechnology in scientific and technological literacy. *Science Education International, 21*(3), 160–175. Retrieved September 26, 2017, from https://eric.ed.gov/?id=EJ904866.

LeBlanc, S., Renninger, S., & Shittu, E. (2016a). Nanotechnology fellows program: Preparing undergraduate students for careers in nanotechnology. In *Proceedings of ASEE conference & exposition* (pp. 1–8). New Orleans, LA.

LeBlanc, S., Sorger, V., & Shittu, E. (2016b). Nanotechnology fellows program: An interdisciplinary practicum for nanotechnology undergraduate education. In *Proceedings of the NSF-AAAS symposium on envisioning the future of undergraduate STEM education: Research and practice*. Washington, DC.

Lee, C.-J., & Scheufele, D. A. (2006). The influence of knowledge and deference toward scientific authority: A media effects model for public attitudes toward nanotechnology. *Journalism & Mass Communication Quarterly, 83*(4), 819–834. https://doi.org/10.1177/107769900608300406.

Lee, C.-J., Scheufele, D. A., & Lewenstein, B. (2005). Public attitudes toward emerging technologies. *Science Communication, 27*(2), 240–267. https://doi.org/10.1177/1075547005281474.

Maienschein, J., & Students. (1999). Commentary: To the future—Arguments for scientific literacy. *Science Communication, 21*(1), 75–87. https://doi.org/10.1177/1075547099021001003.

Malsch, I. (2008). Nano-education from a European perspective. *Journal of Physics: Conference Series, 100*(3), 032001. https://doi.org/10.1088/1742-6596/100/3/032001.

McREL: Education and Public Outreach: NanoLeap: Home Page (n.d.). Retrieved September 30, 2017, from http://www2.mcrel.org/nanoleap/.

National Informal STEM Education Network. (n.d.) *NISE network*. Retrieved September 30, 2017, from http://www.nisenet.org/search/product_type/evaluation-and-research-32.

National Nanotechnology Initiative. (2016). *2016 NNI strategic plan*. Retrieved September 30, 2017, from https://www.nano.gov/2016StrategicPlan.

Porter, A. L., & Youtie, J. (2009). How interdisciplinary is nanotechnology? *Journal of Nanoparticle Research, 11*(5), 1023–1041. https://doi.org/10.1007/s11051-009-9607-0.

Porter, L. A. (2007). Chemical nanotechnology: A liberal arts approach to a basic course in emerging interdisciplinary science and technology. *Journal of Chemical Education, 84*(2), 259. https://doi.org/10.1021/ed084p259.

Roco, M. C. (2002). Nanotechnology a frontier for engineering education. *International Journal of Engineering Education, 18*(5), 488–497.

Roco, M. C. (2003a). Broader societal issues of nanotechnology. *Journal of Nanoparticle Research, 5*(3/4), 181–189. https://doi.org/10.1023/A:1025548512438.

Roco, M. C. (2003b). Converging science and technology at the nanoscale: Opportunities for education and training. *Nature Biotechnology, 21*(10), 1247–1249. https://doi.org/10.1038/nbt1003-1247.

Scheufele, D. A., & Lewenstein, B. V. (2005). The public and nanotechnology: How citizens make sense of emerging technologies. *Journal of Nanoparticle Research, 7*(6), 659–667. https://doi.org/10.1007/s11051-005-7526-2.

Schönborn, K. J., et al. (2014). Development of an interactive immersion environment for engendering understanding about nanotechnology. *International Journal of Virtual and Personal Learning Environments, 5*(2), 40–56. https://doi.org/10.4018/ijvple.2014040104.

Stevens, S., Sutherland, L., & Krajcik, J. (2009). *The big ideas of nanoscale science and engineering: A Guidebook for secondary teachers*. Arlington, VA: National Science Teachers Association. https://doi.org/10.2505/9781935155072.

Wang, H., et al. (2013). Development and evaluation of nanotechnology courses at The George Washington University. *Journal of Nano Education, 5*(1), 79–84. https://doi.org/10.1166/jne.2013.1036.

Chapter 3
Nanotechnology and Education

Daphne L. Schmidt

Call to the Future

Standing at a wooden podium in a lecture hall at CalTech with a blackboard at his back, Richard Feynman changed the way we see the world. It has been nearly 60 years since he presented his lecture, "There's Plenty of Room at the Bottom" to the American Physical Society. Through a series of "what ifs" and "imagine thats" and even throwing in a wager or two, Feynman sparked the imagination of the scientific community with the prospect of manipulating individual atoms and molecules (Feynman 1960). Could it actually be possible to manipulate matter at the nanoscale, creating functional structures that would revolutionize medicine, computer science, manufacturing, and more? In his vision for what the future might hold, he highlighted the need to engage our youth, suggesting a high school competition that would stimulate interest in the nascent field of nanotechnology.

Feynman speculated that there would be two key challenges facing further exploration at the nanoscale. One would be developing precision tools that could facilitate imaging and manipulating atoms and molecules and the other would be dealing with scaling issues such as relative force dominance and interactions. In the years since that lecture, scientists and engineers around the world have taken on Feynman's challenges and the field of nanotechnology has rapidly moved to the forefront of global research. Engaging our youth has taken a bit longer to get started, as will be seen further in this discussion.

Advances in imaging tools in the 1980s represented a tremendous breakthrough for nanotechnology. Suddenly, researchers could "see the unseen." IBM researchers Gerd Binnig and Heinrich Rohrer developed the first electron tunneling microscope in 1981 (which earned them the Nobel Prize in Physics in 1986). Shortly thereafter,

D. L. Schmidt (✉)
Science Education Consultant, Glen Allen, VA, USA

© Springer Nature Switzerland AG 2020
P. M. Norris, L. E. Friedersdorf (eds.), *Women in Nanotechnology*, Women in Engineering and Science, https://doi.org/10.1007/978-3-030-19951-7_3

Fig. 3.1. IBM logo
courtesy of IBM library

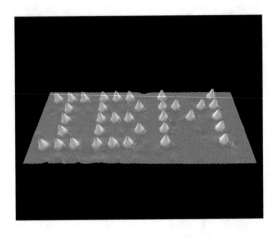

Binnig developed the atomic force microscope, considered a foundational tool for imaging, measuring, and manipulating matter at the nanoscale (Baird et al. 2004). These tools allowed Don Eigler, an IBM researcher, to create one of the most famous images in nanotechnology.

Using a scanning tunneling microscope, he very carefully manipulated 35 xenon atoms to create the IBM logo, a gift to the corporation that gave him a job (Baird et al. 2004) (Fig. 3.1.).

Feynman's vision for the potential of nanotechnology was further expanded as the buckminsterfullerene was discovered by Harry Kroto, Richard Smalley, and Robert Curl in 1985, named for its similarity in appearance to the geodesic dome designed by architect Buckminster Fuller. Additionally, the carbon nanotube, a tubular structure of carbon atom sheets, was discovered by Sumio Iijima in 1991 (Baird et al. 2004). As researchers explored the nanoscale with these new tools, it became quite apparent that nanoscience, by its very nature, lies at the convergence of all scientific disciplines. Both of these carbon allotropes have since served as the basis for a myriad of innovations in materials science, medicine, electronics, and energy storage and manipulation and will no doubt continue to do so in the future.

Richard Feynman stressed the need for our youth to understand the importance of nanotechnology as he recognized that students represent not only the future consumers of technological innovations, but also the future leaders in their research and development. As nanotechnology has advanced over the last 30 years, educators have recognized this call to the future, yet have grappled with how best to engage youth in its often abstract mysteries.

A key challenge for nanotechnology development is the education and training of a new generation of skilled workers in the multidisciplinary perspectives necessary for rapid progress of the new technology. The concepts at the nanoscale (atomic, molecular and supermolecular levels) should penetrate the education system in the next decade in a manner similar to the way the microscopic approach made inroads in the last 50 years (Roco 2002).

Student Readiness: Introducing the World of the Very Small

Understanding student readiness and the process by which educators can help students bridge knowledge is key to the successful integration of nanotechnology in both formal and informal educational settings. Learning theories abound discussing the manner in which humans construct knowledge, but most agree that knowledge is built in varying degrees by biology, environment, and experience (Lutz and Huitt 2004).

Jean Piaget linked readiness to learn to clearly define biological constructs (Piaget 2001). He postulated that children move through a series of developmental stages loosely aligned with chronological age. Each stage presents a unique circumscribed process for receiving and processing information. Children in the sensorimotor stage (ages birth–2 years) learn primarily through trial and error, building an understanding of object permanence. They are not able to understand symbols as representing reality. In the preoperational stage (ages 2–7), children are very egocentric, but begin to ascribe meaning to symbols and develop a wicked imagination! The concrete operational stage (ages 7–11) introduces children to cognitive processing of symbols in a logical manner. They begin to problem solve and develop ideas in their heads, but their work is based on an understanding of concrete objects. In the formal operation stage (ages 11–adulthood), children are able to understand abstract concepts through symbols and solve complex problems.

Although very focused on biological constructs, Piaget also described "schemas," building blocks of knowledge that could be added to and changed through environmental interactions and hands-on experiences. These schemas assist children as they encounter new types of information, facilitating processing of that information by guiding how they respond based on their developmental stage and past experiences (Piaget 2001).

In order for successful integration of nanoscience in educational curricula to be achieved, educators should note that children in the early developmental stages will have difficulty working with abstract concepts. Learners will need some type of scaffolding through environmental interactions and experience to slowly build understanding. Educators must understand the schema that their students are using to effectively help them learn and integrate new knowledge. Lutz and Huitt describe this as developing a spiral curriculum with concepts being "addressed at ever increasing levels of abstractness and complexity" (Lutz and Huitt 2004).

Constructivism approaches learning through the lens of mental structures which become more elaborate and sophisticated through interaction and experience (Bruner 1987). Learners construct knowledge based on knowledge that they already possess. This approach highlights the importance of bridging new knowledge with past experiences, noting the power of hands-on activities to make connections and how linking abstract and concrete concepts can help learners bridge the gap. A common application of this in nanoscience education is playing a size and scale card game where students align familiar objects such as a human hair or a fly to their measurement in nanometers. By using a concrete example, students are able to begin imagining the true size and scale of the nanoworld.

Jerome Bruner believes that certain aspects of any content or principle can be successfully taught to any child no matter the complexity of the topic. It may simply be necessary to revisit the topic as the learner gains more knowledge and complex thought processing abilities (Bruner 1987). For instance, according to the Next Generation Science Standards, students in grade 2 should be able to describe and classify materials according to their observable properties such as color. In grade 5, students explore matter and its interactions including the nature of color as the result of light interacting with the matter being observed (NGSS 2013). By the time students are in grades 7–12, they are exploring color through the complexities of wavelengths of light being reflected and absorbed and the size of the matter being observed. Educators can introduce the idea that the size of an object might impact how it interacts with the world at a very early age, weaving it into the conversations about color and revisiting it as the students build knowledge. By doing so, when students begin manipulating matter at the nanoscale in their physics classrooms, it will make perfect sense to them that gold is yellow at the macroscale and turns red at the nanoscale since they understand the relationship between particle size and its interactions with the world around it.

Equally important is the need to ascribe meaning to learning, providing the "so what" to student inquiry. Meaning allows learners to more easily categorize new information so that it can be added to prior knowledge and utilized. It also adds a powerful motivational factor for learning. For instance, a group of seventh and eighth graders demonstrated a high level of interest in a photolithography lab that modeled the process with "Nanosmores." They could relate the new information to fond memories of camping trips, or backyard adventures. The students understood clearly that this was a layering process, with different layers having different properties lending themselves to the proper function of the model. By establishing a comfort zone of understanding, the students were then able to confidently explore "writing with light," a tremendous advance in nanotechnology. And the fact that they could eat the lab at the end certainly didn't hurt!

Meaningful learning requires understanding the big picture as well as the details (Brooks and Brooks 2000). Nanotechnology is a powerful tool for providing meaningful learning experiences because as learners explore matter at the macroscale all the way down to the nanoscale, they become attuned to how and why matter acts as it does. Thinking becomes more flexible and holistic. Learners are able to give meaning to what is observed, or at least construct the questions that will lead them to answers. As mentioned earlier, the conversation can start small, perhaps asking the question, "Why do sugar granules dissolve faster than a sugar cube?" But building on this, older students might be challenged with why the latest Boeing or Airbus jet is being constructed using a composite material embedded with carbon nanotubes (Chu 2016). They may not know the answer immediately, but they can construct the questions. How will these tiny structures impact the strength of a massive jet? Might there be some safety considerations such as conductivity of electrical discharges as jets pass through powerful storms?

Nanotechnology in the Classroom

As the twenty-first century dawned, it became quite evident that efforts to bring nanotechnology to the classroom needed to be stepped up. University students had been working in the lab on this technology with researchers for at least a decade, but very little nanoscience or nanotechnology was being introduced to primary and secondary students. In fact, nearly 50 years after Richard Feynman's famous lecture, not a single state included vocabulary that specifically referenced nanoscience or nanotechnology in their state standards. So in 2006, Dr. Julia Cothron, former Executive Director of the MathScience Innovation Center (MSiC) in Richmond, Virginia, decided to rectify the situation for the students of Virginia. The MathScience Innovation Center is a STEM hub in Central Virginia supporting 12 school divisions with curricula, teacher professional development, and face-to-face and virtual learning experiences with students K-12 in science, technology, engineering, and math (STEM).

Dr. Cothron: "Approaching its fiftieth anniversary, the MathScience Innovation Center (MSiC) asked the question: How should we help prepare school divisions for the 21st century? Already, the Center led Central Virginia's school divisions in the future-oriented areas of space exploration, integrated Earth systems, and robotics. However, the K-12 curriculum did not address mathematical modeling, engineering, or the behavior of matter at very small scales (nanoscience). Of its future-oriented initiatives, nanoscience was the most important because educators did not understand the scientific principles operating at this level or realize the amazing ways that nanotechnology was revolutionizing all pure and applied sciences, including the development of many common products. Nanoscience was an integrating topic that could be woven through all STEM areas and that enabled the MSiC to Imagine, Create, and Lead. Over multiple years, the MSiC imagined ways that nanoscience could be incorporated into Virginia's educational standards, created and taught model lessons and student courses, and led through teacher training and advocacy at the regional and statewide level."

In early 2006, Dr. Cothron and colleagues at the MSiC began an in-depth study of nanoscience and nanotechnology, with the tremendous support of several universities, including the University of Wisconsin, Georgia Tech, and Penn State, who were beginning to conduct teacher outreach workshops. After attending these workshops and building a team of educators intent on bringing nanoscience and nanotechnology to Virginia's students, the MSiC recognized the importance of its inclusion in state standards. Many months were spent developing language that integrated nanoscience in several key standards. Virginia was in the process of revising its 2003 Science Standards for release in 2010. Dr. Cothron et al. submitted their suggestions for nanoscience inclusion to the Virginia Department of Education. While not all of the suggestions were included, enough were integrated in the standards that a pathway was illuminated. In 2010, Virginia became the first state in the nation to include nanoscience and nanotechnology in its state standards.

Successful integration of nanotechnology into curricular structures requires precise, intentional alignment with current curricula so that the content is perceived as "integral to" and not "in addition to" the content that is currently being taught. At the MathScience Innovation Center, a highly structured curriculum framework, closely aligned with Virginia state standards, was developed based on *The Big Ideas of Nanoscale Science & Engineering,* a publication of the National Science Teachers Association (Stevens et al. 2009). These big ideas were identified by a group of experts including scientists and engineers working at the nanoscale in biology, chemistry, and physics, learning scientists, and science education leaders (both formal and informal). Each big idea was carefully defined with a discussion as to why it should be considered a big idea. In addition, each big idea was explicitly related to secondary curricula and current applications in science and industry. This is vitally important as students construct knowledge about futuristic concepts. As noted by Jerome Ernst, "Curricular activities that incorporate real world examples can enhance students' attitudes about science and emerging ideas" (Ernst 2009). Learning goals for each big idea were also identified targeting specific prerequisite knowledge, potential student difficulties and misconceptions, and what could be expected as learning outcomes (Stevens et al. 2009).

The *Big Ideas of Nanoscale Science & Engineering* include:

- Size and scale
- Structure of matter
- Forces and interactions
- Quantum effects
- Size-dependent properties
- Self-assemble
- Tools and instrumentation
- Models and simulations
- Science, technology, and society

Although developed as a guidebook for secondary teachers, grades 7–12, the concepts included in *The Big Ideas of Nanoscale Science & Engineering* are adaptable for younger learners based on previously discussed parameters.

Nanotechnology in the classroom presents educators with a natural vehicle to explore the interdisciplinary nature of STEM (science, technology, engineering, and math) education. As a convergent discipline, nanotechnology helps learners build integrated knowledge structures that enhance scientific literacy and support creative problem-solving. The move towards STEM education was spurred by the recognition that the manner in which our youth experience education should be more closely aligned to how they will work in their careers. Collaboration and the ability to work across and between different disciplines are crucial to innovation and sustainment. Whether students want to pursue a career in health sciences, high-tech manufacturing, the beauty industry, or design, they will need to possess the cognitive flexibility to see the big picture as well as the details of their expertise. In her article, *For Integrating STEM, Experts Recommend Teaching Nanoscience*, Sarah Sparks contends that although it is a challenge to break down the silos of scientific and mathematical disciplines in a STEM classroom, nanoscience and nanotechnology can ease the way by literally "getting down to the basics" and then building everything back up (Sparks 2015). The very act of observing matter in its smallest parts reveals the interconnectedness of STEM fields and how each discipline together can help build understanding.

As educators are exploring nanotechnology with their students, they are experiencing both challenges and affirming student responses to an integrated curricula. It was reported in one physics classroom that the students responded best to topics regarding the behavior of light. They were intrigued by the nature of light and its behavior at the nanoscale, particularly in discussions about iridescence and thin films. Not surprisingly, given our understanding of how students construct knowledge, they thoroughly enjoyed investigations about carbon nanostructures and hydrophilic/hydrophobic molecules because they could build on understandings that they acquired in their chemistry and biology classes. The properties of carbon and its allotropes "made sense" holistically based on prior knowledge.

One point of frustration was the student's inability to use some of the tools such as AFMs and STMs in their classroom. They were fascinated by how they worked and really wanted to use them in research. Fortunately, universities have now begun to share their resources through such programs as Remote Access Instruments for Nanotechnology (RAIN). Jared Ashcroft of the NACK Network sees RAIN as a vital connection between our future STEM workforce and the research facilities and technologies that will guide their professional development. This program is completely free of charge, bringing advanced technologies to a diverse group of K-12 and college students right in their classrooms. Nineteen universities currently participate in the program (Ashcroft 2018).

Mickey Mouse® gloves were a hit in another classroom exploring the tools and instrumentation that are required to conduct nanotechnological research. Students in pairs were first challenged to build a LEGO® copy of a small model. One student held a timer and the other students assembled the model copy. Then, as if in a magical shrinking machine, students switched roles and the "assembler" was required to do the task again, except this time they must wear a pair of giant Mickey Mouse® gloves. It became very evident how clumsy it felt to work with very small objects without the availability of appropriate tools. This created a great deal of appreciation for the importance of instrument innovation in research and the historical significance of the development of scanning tunneling microscopes in the 1980s. In addition, the experience scaffolded student understanding of size and scale through a hands-on activity that could be used with a broad range of ages.

Most children by middle and high school have been touched in some way by chronic disease, either through family members, friends, or perhaps experiencing it themselves. Nanomedicine has taken the scientific and health community by storm, particularly in health screening and battling cancer. Although chemotherapy has saved many lives by stopping cancer cells in their tracks, it is a systemic treatment, attacking all fast-growing cells in the body including the healthy ones. This can result in terrible side effects, weakening the body's defenses. Recent research has led scientists to find ways to functionalize fullerenes (buckyballs) so that cancer cells are targeted directly, leaving healthy tissue untouched. A group of middle school students were fascinated with an activity using Kool-Aid© cups and gelatin to model a nanoshell, functionalized buckyballs that can selectively target, heat, and destroy cancerous tissues without harming the surrounding tissue. "Researchers are coating gold nanoshells with antibodies and injecting them into the body. The nanoshells circulate in the blood until they attach to antigens on cancer cells. When a laser is shown on the cancerous area, the gold nanoshells heat up—essentially cooking the cancer while the surrounding healthy cells are unharmed" (Jones et al. 2007). The main motivator for the students was how closely this model activity related to what was actually going on in current research. They were also very excited about how soon such technologies would be available on the market for patient therapies. This activity both educated and inspired students to further explore the impact of nanotechnology on society. It also led to some very interesting ethics questions concerning what would happen to the nanoshells once they had destroyed the cancer cells. Would they end up in our rivers and streams? How would that impact wildlife?

In an elementary setting, a fourth-grade educator focused on expanding her students' understanding of size and scale. She shared with them not only how small nanoparticles are, but also how size impacts how matter relates to forces and interactions. The students enjoyed exploring what a nanometer was when they were challenged to cut a paper strip down as small as they could or all the way down to nanoscale, whichever came first. They soon realized that nanosize is VERY small and they did not have the right tools to reach that scale (MRSEC Education Group 2007). Next, the students were introduced to how tiny some very familiar molecules are through an activity that she called "Smellervision." Introduced by the NISE Network, this activity allows students to experience nanoparticles with their senses. Different extracts are placed in balloons and the students try to identify their smell. The students can't see the extract molecules, but they can certainly smell them. What really startled the students was when they realized that the nanoscale scent molecules actually penetrated the membrane of the balloon (NISE Network 2010). And who wouldn't want to explore nanotechnology with rockets! These elementary students were all eyes and ears as they discovered that particle size affects how particles will react in a chemical reaction. Using simple tools such as Alka-Seltzer® tabs and old film canisters, students experienced first hand how size variables impacted the rate that Alka-Seltzer® tabs released gas and "fueled" their rockets. This was then related to surface area, deepening student understanding of chemical reactions.

As every educator knows, education is most impactful on students' lives when they can relate their studies to their own experiences and the world around them. Headlines abound about sunscreens, their proper use, and impact on the environment, particularly coral reefs. Recently, Hawaii passed a bill banning the use of sunscreens containing the chemicals oxybenzone and octinoxate, chemicals that scientists have found to contribute to coral bleaching when washed off in the ocean (The New York Times 2018). We encourage youth to use sunscreen to protect against cancer-causing UV light and the aging effects of the sun, but how can this be accomplished without harming the environment? Students in an exploratory high school course found that sunscreens containing nano-zinc not only seem to be non-harmful to coral reefs, but are also transparent. This appealed to many students who would prefer to avoid the typical "white nose" of lifeguards. By comparing different sunscreens (both macro- and nanomolecule based), and using UV beads as an indicator, students found evidence supporting their hypothesis that nanoparticles in sunscreens are highly effective.

Another group of students were fascinated with what makes matter hydrophobic or hydrophilic. By exploring the structure of matter and hydrogen bonding through a nanotechnology lens, they understood what was going on when they encountered sand that had been nano-engineered to repel water, a product known as Magic Sand®. In a simple water submersion activity, two plastic spoons had sand glued to their surfaces, one coated with regular sand and another coated with Magic Sand® (NISE Network 2011). The educator noted that the Magic Sand® had a nanoscale hydrophobic coating on the grains of sand. Following their experimentation, the students engaged in some collaborative discussions, and were able to imagine numerous creative uses for the nano-engineered sand as well as make predictions about how its use could benefit society. This one experiment had the students experiencing a pure science phenomenon, applying it to current situations, and evaluating its usefulness to society, all through the lens of nanotechnology.

Resources for Educators

There are numerous resources for educators to utilize as they begin to integrate nanotechnology into their classroom curricula. Daniel Herr, working at the Joint School of Nanoscience and Nanoengineering at the University of North Carolina, imagines a transdisciplinary educational ecosystem (Herr 2016). This ecosystem would be made up of a complex community of educators, learners, and community partners sharing the common language of nanoscience. Such an environment would greatly enhance the learning experience, building strong foundations of knowledge, accelerating the synthesis of ideas, and nurturing inquiry into further possibilities. The attached appendix includes just a few of the many online resources available for use in both formal and informal educational settings. Resources include tools to enhance the student learning experience as well as professional development opportunities for educators. Additional invaluable resources are to be found through partnerships with industry and community members. The National Aeronautics and Space Administration (NASA) offers robust support for educators, bringing innovative science right into classrooms through activities and speakers. One lucky group of educators in a Nano Fellows program were mesmerized by Dr. Mia Siochi, a NASA Langley materials scientist, who connected the nanoscience concepts they were exploring with the amazing technologies being used by NASA to build the latest airplanes and rockets. As a consequence of their experience, their students were privy to up-to-the-minute advances in nanotechnology. Making connections with local universities and industry can also lead to resources that will enhance the learning experience. Whether it is a virtual walking tour of a tissue engineering lab, or collaborating to present a hands-on workshop on piezoelectronics in a student conference, these partners are eager to reach out to their communities, building a pipeline of learning.

In Conclusion

Introducing a new way of thinking about education is always challenging and the integration of nanotechnology into current educational models does pose a challenge. But with that challenge comes the thrill of innovation and identifying future possibilities. As more and more states come to recognize the importance of nanotechnology for the future of our students, its concepts will be adopted as an integral part of state standards. Educators and students will have the opportunity to explore matter at the nanoscale in a transdisciplinary educational ecosystem, recognizing nanotechnology's potential to profoundly affect society. Standing at a wooden podium in a lecture hall at CalTech with a blackboard at his back, Richard Feynman changed the way we see the world. As educators, exploring nanotechnology side by side with our students, we can too!

Appendix

Nanotechnology Education Resources
Nanoscale Informal Science Education Network: The NISE Network is a national community of researchers and informal science educators dedicated to fostering public awareness, engagement, and understanding of nanoscale science, engineering, and technology. http://www.nisenet.org/
National Center for Learning and Teaching: NCLT is dedicated to developing and offering nanotechnology-specific instructional modules, professional development, and a network of educator communities related to learning and teaching about the nanoscale. http://www.nclt.us
National Science Digital Library: NSDL is an extensive collection of online resources for science, technology, engineering, and mathematics (STEM) education. Using the search feature provides myriad links related to nanoscience and nanotechnology. http://nsdl.org
National Nanotechnology Infrastructure Network: NNIN provides a wide variety of educational outreach that spans the spectrum of K-12 education. Education and outreach components of the NNIN include an online science magazine for upper elementary and middle school students, Nanooze. http://www.mmm.org/mmim_edu.html and http://www.nanooze.org
National Nanotechnology Initiative: NNI provides resources for students and teachers including information about nanotechnology programs from community colleges to PhDs; a description of the growing Nano and Emerging Technologies Student Network; and links to multimedia contests, videos, and animations. Additionally, a searchable database of nanotechnology education resources can be found at nanoHUB.org. https://www.nano.gov/education-training/teacher-resources
Materials World Modules: An NSF-funded program produced a series of interdisciplinary modules based on topics in material science, including composites, ceramics, concrete, biosensors, biodegradable materials, smart sensors, polymers, food packaging, and sports materials. Modules are designed for middle and high school STEM classes. http://www.materialsworldmodules.org
Exploring the Nanoworld: The University of Wisconsin-Madison materials Research Science and Engineering Center (UW MRSEC) uses examples of nanotechnology and advanced materials to explore fundamental science and engineering concepts at the college level and to share the "wow" and potential of these fields with public audiences. The website includes movies, kits, references, and teaching modules for K-12 teachers. http://mrsec.wisc.edu/Edetc/index2.html
Center for Nanotechnology Education and Utilization: CNEU at Penn State's College of Engineering offers resources such as webcasts, video modules, workshops for educators, and resources related to careers in nanotechnology from both the educator's and industry's point of view. http://www.cneu.psu.edu
NanoWerk: Explore one of the world's most comprehensive lists of nanoscience and nanotechnology resources. This site includes the latest news makers in nanotechnology. Http://www.nanowerk.com

Nanotechnology Education Resources
Nanotechnology Applications and Career Knowledge (NACK) Network: Nano4Me.org provides webinar and workshop information for educators, educational resources for students, and guides for developing integrated nanotechnology curricula. http://www.nano4me.org
Understanding Nano: This website is dedicated to providing clear and concise explanations of nanotechnology applications along with information on companies working in each area. www.understandingnano.com
Nanozone: A fun website for students and the general public interested in learning about nanotechnology. http://www.nanozone.org
NanoHUB: An online community of researchers and educators hosting a rapidly growing collection of simulation programs for nanoscale phenomena that run in the cloud and are accessible through a web browser. Also offers workshops, virtual tools, and databases. https://nanoHUB.org
National Science Foundation: The Nanoscience Classroom Resources page provides a diverse collection of lessons and web resources for classroom teachers, their students, and students' families. https://www.nsf.gov/news/classroom/nano.jsp

References

Ashcroft, J. (2018). Remotely accessible instruments in nanotechnology (RAIN) labs. Retrieved January 21, 2019, from https://nanohub.org/resources/27889.

Baird, D., Nordman, A., & Schummer, J. (Eds.). (2004). *Discovering the nanoscale*. Amsterdam: IOS Press.

Brooks, J., & Brooks, M. (2000). *Search of understanding: The case for constructivist classrooms*. Upper Saddle River, NJ: Prentice-Hall.

Bruner, J. (1987). *Actual minds, possible worlds*. Cambridge, MA: Harvard University Press.

Chu, J. (2016). Method to reinforce carbon nanotubes could make airplane frames lighter, more damage-resistant. Retrieved January 21, 2019, from https://phys.org/news/2016-08-method-carbon-nanotubes-airplane-lighter.html.

Eigler, D. M., & Schweizer, E. K. (1990). Positioning single atoms with a scanning tunneling microscope. *Nature, 344*, 524–526.

Ernst, J. V. (2009). Nanotechnology education, contemporary content and approaches. *The Journal of Technology Studies, 35*(1), 3–8.

Feynman, R. P. (1960). There's plenty of room at the bottom. *Engineering and Science, 23*(5), 22–36.

Herr, D. (2016). The need for convergence and emergence in twenty-first century Nano-STEAM+ education ecosystems. In K. Winkelmann & B. Bhushan (Eds.), *Science policy reports* (pp. 83–115). https://doi.org/10.1007/978-3-319-31833-2_3.

Jones, G., Falvo, M., Taylor, A., & Broadwell, B. (2007). *Nanoscale science activities for grades 6–12*. Arlington, VA: NSTA Press.

Lutz, S., & Huitt, W. (2004). Connecting cognitive development and constructivism: Implications from theory for instruction and assessment. *Constructivism in the Human Sciences, 9*(1), 67–90.

MRSEC Education Group. (2007). Cutting it down to nano outreach activity. Retrieved January 28, 2019, from https://education.mrsec.wisc.edu/cutting-it-down-to-nano-outreach-activity/.

Next Generation Science Standards. (2013). Retrieved January 21, 2019, from https://www.next-genscience.org.

NISE Network. (2010). Exploring size-scented balloons. Adapted from Odors Aloft, No Hassle Messy Science with a Wow: Chemistry in the K-8 Classroom published by the Oregon Museum of Science and Industry, 2007. Retrieved January 21, 2019, from http://www.nisenet.org/sites/default/files/catalog/uploads/4130/sizeballoon_guide_15nov10.pdf.

NISE Network. (2011). Magic sand-nanosurfaces. Retrieved January 28, 2019, from http://www.nisenet.org/catalog/magic-sand-nanosurfaces.

Piaget, J. (2001). *The psychology of intelligence* (2nd ed.). London: Routledge. [Originally published in 1950].

Roco, M. C. (2002). Nanoscale science and engineering education activities in the United States. *Journal of Nanoparticle Research, 4*, 271–274.

Stevens, S., Sutherland, L., & Krajcik, J. (2009). *The big ideas of nanoscale science & engineering*. Arlington, VA: NSTA Press.

The New York Times. (2018). Hawaii passes bill banning sunscreen that can harm coral reefs. Retrieved on January 28, 2019 from https://www.nytimes.com/2018/05/03/travel/hawaii-sunscreen-ban.html.

Sparks, S. (2015). For Integrating STEM, Experts Recommend Teaching Nanoscience. Retrieved January 21, 2019, from https://www.edweek.org/ew/articles/2015/01/07/lessons-on-small-particles-yield-big-gains.html.

Chapter 4
Nanotoxicology: Developing a Responsible Technology

Christie M. Sayes

Prologue

Introduction to Nanotechnology

In 2003, Sayes' advisor (Vicki Colvin, Rice University) published a perspective piece in Nature Biotechnology (Colvin 2003). Colvin noted that there was a mounting public discussion surrounding the environmental and social costs of nanotechnology versus its many benefits. At the time, there was speculation that engineered nanomaterials were being incorporated into commercial and industrial products (Maynard et al. 2006). In fact, by 2006, more than 300 products on the market claimed to be nano-enabled (Maynard et al. 2006). According to WikiBooks (https://en.wikibooks.org/wiki/Nanotechnology/Glossary), the term *"nano-enabled is used to refer to devices or systems that utilize some aspect of nanotechnology to enhance their function. Products that act solely on the macroscale but have some enhancement due to nanotechnology are sold as being nano-enabled."* The debate that ensued in the early 2000s resulted in a request for more information about the toxicological and environmental effects of direct and indirect exposure to nanomaterials (Foss Hansen et al. 2008; Maynard et al. 2006). Scientists, regulators, and consumers alike were concerned about the lack of clear guidelines to quantify the potential adverse health effects induced by engineered nanomaterials.

C. M. Sayes (✉)
Department of Environment Science, Baylor University, Waco, TX, USA
e-mail: christie_sayes@baylor.edu

© Springer Nature Switzerland AG 2020
P. M. Norris, L. E. Friedersdorf (eds.), *Women in Nanotechnology*, Women in Engineering and Science, https://doi.org/10.1007/978-3-030-19951-7_4

The State of Nanotoxicological Sciences

This book chapter reviews critical research related to the assessment of the environmental health impact of engineered nanomaterials. The literature in this specific area of nanotechnology is extensive, but there are certain key messages that are emphasized time and time again. First, it cannot be assumed that all nanomaterials are the same as their bulk counterparts; but it also cannot be assumed that nanomaterials are more toxic (Warheit 2008; Sayes and Warheit 2009). Second, much of the discussion in the nanotoxicology literature is based on extrapolations from better understood molecular species or larger particulate matter (Oberdörster et al. 2005b). Third, formal risk assessment for engineered nanomaterials (as a broad class of substances) is not possible (Warheit et al. 2009a; Warheit 2008; Borm et al. 2006; Bartlett et al. 2015).

The main components of a risk analysis are useful and offer a decent structure to the content herein. This chapter reviews nanomaterial exposure (inclusive of exposure routes, quantifying concentration, and considering particle transformations) along with potentially induced adverse health effects (descriptive observations versus mechanisms of action in in vitro and in vivo biological test systems). Both issues are equally important when characterizing the biological and environmental risks of nanomaterials (Warheit 2008).

A Concise Literature Review

Nanomaterial Exposure

Exposure to engineered nanomaterials is a critical component when assessing the risks associated with nanotechnology (Wijnhoven et al. 2009; Robichaud et al. 2009). To date, researchers have identified three broad types of scenarios where exposure to nano-entities may occur: occupational, consumer, or environmental. Occupational exposure occurs in the workplace and can account for either direct exposure, such as handling nano-powders in a supply chain, or indirect exposure, such as inhaling solid particle aerosols in workplace common spaces. Consumer exposure describes a vast range of scenarios, including but not limited to, ingestion of nano-additives from food, intravenous injection of nano-drugs, or dermal penetration from cosmetics. Lastly, environmental exposure is primarily concerned with nano-waste, i.e., industrial debris laced with nanoparticle starting material, destruction of nano-intermediates that failed quality control specifications, or disposal of nano-enabled products at the end of a product's life cycle.

Occupational exposure to engineered nanoparticles, and their associated adverse health effects, has been a critical area of research for the National Institute for Occupational Safety and Health (NIOSH) for over a decade. The institute has produced a multitude of scientific research papers and other educational literature primarily focused on studying the potential worker health risks due to the manufacture and use of nanomaterials. To date, the NIOSH studies have indicated that *"low solubility nanoparticles are more toxic than larger particles on a mass for mass basis* (https://www.cdc.gov/niosh/topics/nanotech/default.html)." Furthermore, *"there are strong indications that particle surface area and surface chemistry are responsible for observed responses in cell cultures and animals. Studies suggests that some nanoparticles can move from the respiratory system to other organs. Research is continuing to understand how these unique properties may lead to specific health effects."* Hyperlinks to NIOSH's guidance and publication on nanotechnology environmental, health, and safety can be found here: https://www.cdc.gov/niosh/topics/nanotech/pubs.html.

There are a few central components of an exposure assessment needed to adequately assess the risk of engineering nanomaterials (Wijnhoven et al. 2009; Christensen et al. 2011; Powers et al. 2013). First, the assessor must describe the material and its potential applications. Second, a comprehensive report of the material's physicochemical properties must be compiled. Lastly, the assessment must include a compilation of intentional and unintentional routes of possible transport through an occupational, consumer, or environmental scenario.

Metrology

When measuring the physicochemical characteristics of nanomaterials, each property should be measured using the most appropriate technique(s) (Nel et al. 2009). Research has shown that the most valuable material characterization reports are those whose initial results were confirmed against a control material, reference material, or using an orthogonal analytical technique. No single technique can accurately describe the physicochemical properties of a nanomaterial. Issues such as method limitations, nontrivial sample preparation, and addition of appropriate controls are considered when analyzing nanomaterial samples. The physical and chemical measurements of the nanomaterial are also affected by solvent type, route of exposure, and other environmental conditions (e.g., pH, temperature, humidity, or extra substituents) (Hardman 2006). The successful development of safe nanomaterials requires a strong collaborative effort among analytical chemists and material scientists with product developers.

Bridging the fields of chemists, material scientists, and product developers requires collaboration (Lujan and Sayes 2017). Thoughtful design of the synthesis and characterization methods used for nanomaterial systems is imperative when assessing exposure and risk. The current recommended guidelines for reporting the minimal information needed to characterize nanomaterial systems for risk characterization include:

1. Size, surface area, and structure (obtained from electron microscopy, when possible)
2. Identity of small-molecular-weight contaminants (such as residual solvents or metals intercalated within the nanomaterial's structure)

3. Manufacturing procedure (from starting material to formulation to final product)
4. Information about surface functional groups, coatings, or charge
5. Transformations of any of the aforementioned properties during sample preparation, handling, or administration in final form

The Most Is Known About Aerosolized Nanomaterials

When studying the publicly available nanotoxicology literature, it is clear that occupational exposure to aerosolized nanomaterials is the most highly investigated area (Sayes et al., 2007b; Sayes and Warheit 2009). While dermal, ingestion, and intravenous exposures to nanomaterials do occur and are a concern to environmental health professionals, the exposure data clearly shows that the most likely route of nanoparticle exposure is inhalation in the workplace. Therefore, studies focused on detecting, measuring, and mitigating particles in the breathing zone were (and probably still are) appropriately important. Available air sampling area monitoring and personal exposure devices were effective and practical—there seemed to be little need to develop new equipment for the detection of aerosolized nanoparticles (Ramachandran et al. 2011; Berg et al. 2010). However, as electron microscopy became a routinely used material characterization tool, research showed that some engineered particles (such as those less than 30 nm) penetrated through air filters and were inhaled or ingested by workers in small concentrations (Tsai et al. 2011, 2009). The solution to this issue was the implementation of engineering controls along with personal protection equipment, which demonstrated efficient containment and further mitigated exposure to airborne nanoparticulate matter. Analysis of nanomaterials in waste disposal was more challenging and remains an active area of research for scientists and engineers alike. Today, nanomaterial facilities use a combination of engineering controls, personal protective equipment (PPE), and standard operating procedures (SOP) for preparation, operation, and disposal.

Recent Advances in Physicochemical Characterization

The features of engineered nanomaterials are exceptionally varied to the point where even particles of the same composition behave differently based simply on differing size, shape, surface charge, or crystallographic structure. As the risk of exposure to this diverse class of materials increases, the demand for integrated approaches to analyze the interface between the nanomaterial and potential biological or environmental receptor is needed (Sayes and Lujan 2017). To that end, advances in the field of analytical microscopy allow scientist to view the "nano-bio interface" in greater resolution and magnification than ever before (Alkilany et al. 2013). For example, electron microscopies can be harnessed to create two-dimensional images of cells exposed to engineered nanomaterials; atomic force microscopy can produce a three-dimensional image based on small amount of force used to measure the height of the specimen (Sayes and Lujan 2017). Microscopy has been, and still is, the most heavily relied upon tool to characterize the unique properties of engineered nanomaterials. This tool provides critical information that can be used to understand the nano-bio interface in greater detail.

Environmental Transformations

When engineered nanomaterials are produced, they exist in a colloidal and nonionic form (Adegboyega et al. 2016; Sharma et al. 2017). However, research has shown that this initial form isn't always stable; many particle types have been shown to transform when incubated in physiological conditions (i.e., serum-rich biofluids) or in environmental conditions (i.e., natural organic matter (NOM) and/or UV light). Engineered materials on the nanometer size scale can transform from single-particle state to their ionic state, agglomerated form, or both (Levard et al. 2012). Researchers have also hypothesized that these transformed states may be even more environmentally damaging compared to the original forms due to the increase in surface area when the particles break down, introduction of ions into the surrounding environment, and an increase in reactivity with adjacent chemicals or biomolecules (Sharma et al. 2014). To date, most nanotoxicology studies report the effects of pristine nanomaterials on a biological or ecological test system; future research should focus on the effects of transformed particles on the same types of test systems.

More on Physiological Transformations

Upon entering the bloodstream, proteins readily adsorb onto and desorb from the surface of an engineered nanomaterial. This is a critical physicochemical property of nanoparticles used in drug product formulations (Lesniak et al. 2012). The cloud of adsorbed proteins is known as the "protein corona" and is postulated to alter either the drug product's therapeutic (positive) or the toxicological (negative) effects (Lundqvist et al. 2008). In order to understand the effects related to the protein corona, detection and identification of the proteins which make up the corona are necessary. In addition to protein corona formation, nanoparticles in systemic circulation either disintegrate over time or accumulate in certain organs (Casals et al. 2010). The metabolites of nanoparticles are an active area of basic biomedical research, with only a few examples of intact materials being readily excreted.

Relevance of Nanotoxicology to Nanomedicine

The increasing use of nanomaterials in the preparation of pharmaceuticals requires both manufacturing and analytical considerations in order to establish the best set of quality metrics suitable for drug product performance as well as risk assessment (Lujan and Sayes 2017; Sayes 2014). A range of different nanomaterial systems exist including crystallized active pharmaceutical ingredients (API), additives, and carriers.

These formulations often require more complex production and characterization strategies than conventional dosage forms. The advantage of using nano-enabled systems in pharmaceutical science is that the effective and desired function of the drug product can be thoughtfully designed and controlled through modern manufacturing processes. Analytical considerations of drugs, additives, and carriers (as well as the manner in which they are measured) are connected to quality control. Ultimately, the objective of incorporating a nanotoxicological framework into a nanomedicine study is to consider the entire drug product life cycle with respect to its manufacture, storage, use, and eventual fate. Most researchers are of the opinion that the tools and approaches to address the needs of these products exist; however, more research may be needed for each unique formulation and application (Oberdörster et al. 2005a). Recent advances have been made to hasten the safety testing of nanomaterial drug products and their use as diagnostic or therapeutic agents (Kim et al. 2010). The success of nano-enabled agents will be defined by many factors and will require a multidisciplinary strategy to achieve the best possible result. Critical factors that help boost technological advances of nanomaterial drug products include:

1. Increasing prospect to commercialize the products
2. Educating new inventors and investors about the nuances in patentable technologies
3. Developing an agreed-upon nanomedicine roadmap to serve as guidance for all stakeholders
4. Permanent quest to develop new materials to address the greatest biomedical challenges

Nanomaterial Health Effects

It is important to obtain as much information about nanomaterials for both exposure and hazard assessments (Warheit 2008). Just as no single technique can accurately describe all relevant physical and chemical properties of a material, multiple methods are needed to accurately predict human and environmental health effects (Berg et al. 2009). Nanomaterial characterization is of critical importance when studying nanotoxicology. In addition, characterization of the biological test system is required to describe both descriptive and mechanistic effects.

Descriptive Toxicology Observations Come Before Mechanistic Analyses

Descriptive toxicology is concerned with reporting the dose-response relationship between the concentration of a chemical test substance and the health effects of the biological test system. Most of the early nanotoxicology literature, including Sayes' own contributions to the field, can be described as pioneering work and opened the emerging field of study now known as the environmental, health, and safety (EHS) of advanced materials. Two of the first studies reporting descriptive toxicological effects of engineered nanomaterials were by Derfus et al. (2004) and Sayes et al. (2004) (Derfus et al. 2004; Sayes et al. 2004). Derfus investigated cadmium selenide (CdSe) core quantum dot (QD) cytotoxic responses to primary hepatocytes isolated from rats. Cytotoxicity was caused by a combination of small-size penetrating multiple subcellular organelles and subsequently leaching cadmium ions into the cytosolic space. The team concluded that CdSe QDs can be rendered nontoxic and used to track cell migration when the particle's surface is coated with either a zinc sulfide (ZnS) shell or a bovine serum albumin (BSA) corona.

Sayes et al. studied the differential cytotoxic effects of water-soluble fullerene species in human skin (HDF) and liver carcinoma (HepG2) cells. In both cell types, the lethal dose of fullerenes changed over seven orders of magnitude with relatively minor alterations in fullerene structure. The least surface-derivatized species was substantially more cytotoxic than most surface-derivatized caged structure. The mechanism of toxic action was reported as oxidative damage to the cell membranes. The work highlighted a strategy for tuning the toxicity of a specific engineered nanomaterial for certain applications (such as cancer therapeutics or bactericides). These publications represent two of the most highly cited papers in nano-EHS. Other papers, studying similar toxicity endpoints, soon followed.

The Search for Unique Mechanisms of Action

As the available research published in the peer-reviewed literature grew, the focus of the toxicology studies evolved from simple descriptive observations to complex mechanistic analyses (Romoser et al. 2011, 2012; Berg et al. 2013). Research programs used molecular sciences more frequently (such as gene regulation and protein expression) and nanotoxicology papers started appearing in molecular journals such as *Molecular Immunology*, *Carbon*, and *Free Radical Biology and Medicine* (Guo et al. 2009; Berg et al. 2009, 2010; Romoser et al. 2011, 2012). Molecular reactions begin with nanoparticle extracellular exposure. After particle endocytosis, reactive oxygen species are often generated and initiate cascading events. MAPK, Nrf2, DNA, and mitochondrial damage soon follow with increased ROS production; cell cycle arrest, apoptosis, and inflammatory responses are the most commonly cited adverse health outcomes. Through translocation, suppression, transcription, and phosphorylation processes, individual cells can experience multiple cascading events after nanoparticle exposure (Lujan and Sayes 2017). There is not enough available data to correlate each nanoparticle to one specific adverse reaction; current data suggests that multiple particle types can result in multiple routes to cell damage or death.

Most notably, only a few papers published with a molecular science component consider the effect of the particle's surface functional group on the induced biochemical signaling (Zhang et al. 2015; Sayes et al. 2006). It is reasonable to anticipate that the first biochemical reaction of a biomolecule would be with the surface functional group of the nanoparticle of interest (functional groups are of similar size and weight to a peptide, enzyme, or lipid; the nanoparticle itself is orders of magnitude larger).

The functional groups decorated on the surface of an engineered nanomaterial can range from biocompatible molecular chains, such as polyethylene glycol (PEG), to terminating linear molecules, such as mercaptocarboxylic acid (Sperling and Parak 2010). The type of surface functionalization varies depending on the intended application of the nanomaterial. Some of the most explored biomedical uses of nanoparticle surface functionalization include avoiding the immune system, targeting specific cell types, and improving medical imaging through microvascular circulation (Storm et al. 1995; Huynh et al. 2010; Shenoy et al. 2006).

There are many papers speculating that the nanoparticle surface chemistry might influence the nanomaterial's toxicokinetic properties (Slowing et al. 2006; Mout et al. 2012; Villanueva et al. 2009). Toxicokinetic properties describe the uptake and elimination of the material over time, while toxicodynamic properties emphasize the effects of the nanomaterial on the organism over time (Topuz and Van Gestel 2015). These two physiological properties inform the physicochemical attributes needed in comprehensive toxicological evaluations.

In Vitro Versus In Vivo Data in the Pulmonary Toxicology Literature

When considering the possible induced toxicities after nanomaterial pulmonary exposure, the comparisons of results between in vivo (whole animal inhalation, instillation, or aspiration exposures) versus in vitro (inoculation to lung cells in culture) measurements often demonstrate little correlation (Warheit et al. 2009b; Sayes et al. 2007b). This could be due to the fact that in vivo studies include different types of endpoint analyses than in vitro studies: tissue versus single-cell monolayers, time course of exposure, and gene/protein biomarkers. It remains clear that in vitro cellular systems still need further development, standardization, and validation (relative to human health effects) in order to provide useful screening data on the relative toxicity of engineered nanomaterials.

In studies designed to assess the capacity of in vitro studies to predict the pulmonary toxicity of particles in vivo, the cell culture test systems utilized were not accurate screens for inflammatory potential in the whole-animal model. It was concluded, however, that the in vitro model may be more suitable for mechanistic toxicity studies wherein hypotheses are being tested (Sayes et al. 2004, 2007a).

Other studies, that focus on in vitro–in vivo correlation (IVIVC), have demonstrated acceptable use of cell and tissue cultures as alternative models to animals in toxicological studies (Oberdörster et al. 2005b; Ramachandran et al. 2011). When the model is properly validated and the collected data is verified, in vitro toxicology models can decrease the time needed to develop a nano-enabled formulation, set relevant physicochemical characterization standards, and serve as a surrogate for detailed gene or protein pathway analyses. Researchers in this field of study know the limits of extrapolating in vitro results to whole-animal endpoints. For example, corrections must be made when reporting numerical values of lethality, and assessing bioavailability parameters, differences in postexposure time periods, and body weights among species. It is worth noting that these same corrections must also be implemented when extrapolating data from rodent studies to human health risk assessments (Sayes et al. 2007b).

Risk Science

Often, nanotoxicological studies inform risk science. A risk assessment, defined as a systematic process of evaluating the potential danger, harm, or loss that may be involved in a projected activity or undertaking, is the product of the combined analysis of both hazard and exposure evaluations (Borm et al. 2006; Christensen et al. 2011; Warheit et al. 2007; Wijnhoven et al. 2009). This type of assessment (as well as these types of evaluations) greatly benefits from the inclusion of material characterization, descriptive observations in whole animals, mechanistic analyses from cell or tissue culture, ecological effects to target receptors, and environmental transformations over time. Information gained from hazard evaluations provides a basis for responsible risk management decisions. For a human health risk assessment, justification for the inclusion of data sets rests on the following criteria:

1. Potential route of exposure related to human health effects
2. Screening for potential carcinogenic effects
3. Screening for potential toxic effects in representative aquatic organisms

When performing risk evaluations, it is imperative to (1) review as much data as possible (i.e., material characterization, hazardous effects, and exposure profile) and then (2) compile the information in such a way as to identify the risks that need to be considered based on the nanomaterial and its application (Warheit et al. 2009a; Warheit 2008). Implementing risk management (in the workplace for human health or in the environment for ecology) involves a review of the options to control the risks identified and to develop (or refine) a recommendation for the best risk management scheme for a given material or application. Risk assessments and management schemes are constantly reviewed and updated with the latest information available from science, regulation, and business processes. Ultimately, risk scientists are charged with ensuring that the management system(s) setup is working as expected and adapting as situations change.

Other Advanced Materials and Emerging Contaminants

Most recently, the research interests of the nanotoxicology community have progressed to also include the environmental health effects of advanced materials. Advanced materials are new materials, or modifications to existing materials, that possess superior performance in one or more characteristics that are critical for the application under consideration.

The basic concepts of material characterization and mechanistic analyses that were demonstrated in the nanotoxicology literature are applicable to a wide range of environmental health and safety research questions today (Zeidan et al. 2015; Adegboyega et al. 2016; Sharma et al. 2017).

Epilogue

As recent as 10 years ago, most nanomaterial suppliers were small start-up companies or academic laboratories. Nanomaterials varied in type, purity, and method of synthesis. This variation was exacerbated in the corresponding toxicity results. Therefore, the community began to require the inclusion of physicochemical characterization in manuscripts submitted for publication. The prerequisite set a movement in motion—the effort to link physicochemical properties to overserved toxicological effects. As a result, it becomes increasingly clear that researchers were familiar with pristine particles but lacked characterization knowledge about nanomaterials incorporated into products or complex formulations. When nanotoxicology research first began, much of the physical and chemical characterization data were unavailable or unknown. This was in part because companies that produce nanomaterial final products and intermediate formulations often consider characterization data as proprietary information; the toxicology and environmental fate data available from suppliers was minimal and highly variable. Today, there is a noticeable collaboration among discovery scientists and product developers. A general consensus exists among the scientific and regulatory communities: characterization of pristine nanoparticles, nanomaterial intermediates, and nano-enabled products should be tested and retested in multiple laboratories, both internal and external to the original developer, in an effort to produce both safe and effective nano-enabled products for years to come.

The recommendations of future research need in the nanotoxicology include:

1. Metrology tools that can detect nanoparticles at low concentrations in complex mixtures
2. Toxicology models that can assess potential human health effects in a rapid and inexpensive manner
3. Mathematical or statistical analysis techniques that can build useful quantitative structure-activity relationships for nanomaterials
4. Effective and strategic research programs, partnerships, and sources of funding that bring stakeholders from all communities together

References

Adegboyega, N., Sharma, V., Cizmas, L., & Sayes, C. (2016). UV light induces Ag nanoparticle formation: Roles of natural organic matter, iron, and oxygen. *Environmental Chemistry Letters, 14*, 353–357.

Alkilany, A., Lohse, S., & Murphy, C. (2013). The gold standard: Gold nanoparticle libraries to understand the nano-bio interface. *Accounts of Chemical Research, 46*, 650–661.

Bartlett, J. A., Brewster, M., Brown, P., Cabral-Lilly, D., Cruz, C. N., David, R., Eickhoff, W. M., Haubenreisser, S., Jacobs, A., Malinoski, F., Morefield, E., Nalubola, R., Prud'homme, R. K., Sadrieh, N., Sayes, C. M., Shahbazian, H., Subbarao, N., Tamarkin, L., Tyner, K., Uppoor, R., Whittaker-Caulk, M., & Zamboni, W. (2015). Summary report of PQRI workshop on nanomaterial in drug products: Current experience and management of potential risks. *The AAPS Journal, 17*, 44–64.

Berg, J., Romoser, A., Banerjee, N., Zebda, R., & Sayes, C. (2009). The relationship between pH and zeta potential of similar to 30 nm metal oxide nanoparticle suspensions relevant to in vitro toxicological evaluations. *Nanotoxicology, 3*, 276–283.

Berg, J., Ho, S., Hwang, W., Zebda, R., Cummins, K., Soriaga, M., Taylor, R., Guo, B., & Sayes, C. (2010). Internalization of carbon black and maghemite iron oxide nanoparticle mixtures leads to oxidant production. *Chemical Research in Toxicology, 23*, 1874–1882.

Berg, J. M., Romoser, A. A., Figueroa, D. E., West, C. S., & Sayes, C. M. (2013). Comparative cytological responses of lung epithelial and pleural mesothelial cells following in vitro exposure to nanoscale SiO_2. *Toxicology In Vitro, 27*, 24–33.

Borm, P. J., Robbins, D., Haubold, S., Kuhlbusch, T., Fissan, H., Donaldson, K., Schins, R., Stone, V., Kreyling, W., Lademann, J., Krutmann, J., Warheit, D., & Oberdorster, E. (2006). The potential risks of nanomaterials: A review carried out for ECETOC. *Particle and Fibre Toxicology, 3*, 11.

Casals, E., Pfaller, T., Duschl, A., Oostingh, G. J., & Puntes, V. (2010). Time evolution of the nanoparticle protein corona. *ACS Nano, 4*, 3623–3632.

Christensen, F. M., Johnston, H. J., Stone, V., Aitken, R. J., Hankin, S., Peters, S., & Aschberger, K. (2011). Nano-TiO_2—Feasibility and challenges for human health risk assessment based on open literature. *Nanotoxicology, 5*, 110–124.

Colvin, V. L. (2003). The potential environmental impact of engineered nanomaterials. *Nature Biotechnology, 21*, 1166–1170.

Derfus, A., Chan, W., & Bhatia, S. (2004). Probing the cytotoxicity of semiconductor quantum dots. *Nano Letters, 4*, 11–18.

Foss Hansen, S., Maynard, A., Baun, A., & Tickner, J. A. (2008). Late lessons from early warnings for nanotechnology. *Nature Nanotechnology, 3*, 444–447.

Guo, B., Zebda, R., Drake, S. J., & Sayes, C. M. (2009). Synergistic effect of co-exposure to carbon black and Fe_2O_3 nanoparticles on oxidative stress in cultured lung epithelial cells. *Particle and Fibre Toxicology, 6*, 4.

Hardman, R. (2006). A toxicologic review of quantum dots: Toxicity depends on physicochemical and environmental factors. *Environmental Health Perspectives, 114*, 165–172.

Huynh, N. T., Roger, E., Lautram, N., Benoît, J.-P., & Passirani, C. (2010). The rise and rise of stealth nanocarriers for cancer therapy: Passive versus active targeting. *Nanomedicine, 5*, 1415–1433.

Kim, S. C., Chen, D.-R., Qi, C., Gelein, R. M., Finkelstein, J. N., Elder, A., Bentley, K., Oberdörster, G., & Pui, D. Y. (2010). A nanoparticle dispersion method for in vitro and in vivo nanotoxicity study. *Nanotoxicology, 4*, 42–51.

Lesniak, A., Fenaroli, F., Monopoli, M. P., Åberg, C., Dawson, K. A., & Salvati, A. (2012). Effects of the presence or absence of a protein corona on silica nanoparticle uptake and impact on cells. *ACS Nano, 6*, 5845–5857.

Levard, C., Hotze, E., Lowry, G., & Brown, G. (2012). Environmental transformations of silver nanoparticles: Impact on stability and toxicity. *Environmental Science & Technology, 46*, 6900–6914.

Lujan, H., & Sayes, C. (2017). Cytotoxicological pathways induced after nanoparticle exposure: Studies of oxidative stress at the 'nano-bio' interface. *Toxicology Research, 6*, 580–594.

Lundqvist, M., Stigler, J., Elia, G., Lynch, I., Cedervall, T., & Dawson, K. A. (2008). Nanoparticle size and surface properties determine the protein corona with possible implications for biological impacts. *Proceedings of the National Academy of Sciences, 105*(38), 14265–14270.

Maynard, A. D., Aitken, R. J., Butz, T., Colvin, V., Donaldson, K., Oberdörster, G., Philbert, M. A., Ryan, J., Seaton, A., Stone, V., Tinkle, S. S., Tran, L., Walker, N. J., & Warheit, D. B. (2006). Safe handling of nanotechnology. *Nature, 444*, 267–269.

Mout, R., Moyano, D. F., Rana, S., & Rotello, V. M. (2012). Surface functionalization of nanoparticles for nanomedicine. *Chemical Society Reviews, 41*, 2539–2544.

Nel, A., Madler, L., Velegol, D., Xia, T., Hoek, E., Somasundaran, P., Klaessig, F., Castranova, V., & Thompson, M. (2009). Understanding biophysicochemical interactions at the nano-bio interface. *Nature Materials, 8*, 543–557.

Oberdörster, G., Maynard, A., Donaldson, K., Castranova, V., Fitzpatrick, J., Ausman, K., Carter, J., Karn, B., Kreyling, W., & Lai, D. (2005a). Principles for characterizing the potential human health effects from exposure to nanomaterials: Elements of a screening strategy. *Particle and Fibre Toxicology, 2*, 8.

Oberdörster, G., Oberdörster, E., & Oberdörster, J. (2005b). Nanotoxicology: An emerging discipline evolving from studies of ultrafine particles. *Environmental Health Perspectives, 113*, 823.

Powers, C. M., Bale, A. S., Kraft, A. D., Makris, S. L., Trecki, J., Cowden, J., Hotchkiss, A., & Gillespie, P. A. (2013). Developmental neurotoxicity of engineered nanomaterials: Identifying research needs to support human health risk assessment. *Toxicological Sciences, 134*, 225–242.

Ramachandran, G., Wolf, S. M., Paradise, J., Kuzma, J., Hall, R., Kokkoli, E., & Fatehi, L. (2011). Recommendations for oversight of nanobiotechnology: Dynamic oversight for complex and convergent technology. *Journal of Nanoparticle Research, 13*, 1345–1371.

Robichaud, C. O., Uyar, A. E., Darby, M. R., Zucker, L. G., & Wiesner, M. R. (2009). Estimates of upper bounds and trends in nano-TiO$_2$ production as a basis for exposure assessment. *Environmental Science & Technology, 43*, 4227–4233.

Romoser, A., Chen, P., Berg, J., Seabury, C., Ivanov, I., Criscitiello, M., & Sayes, C. (2011). Quantum dots trigger immunomodulation of the NF kappa B pathway in human skin cells. *Molecular Immunology, 48*, 1349–1359.

Romoser, A., Figueroa, D., Sooresh, A., Scribner, K., Chen, P., Porter, W., Criscitiello, M., & Sayes, C. (2012). Distinct immunomodulatory effects of a panel of nanomaterials in human dermal fibroblasts. *Toxicology Letters, 210*, 293–301.

Sayes, C. (2014). The relationships among structure, activity, and toxicity of engineered nanoparticles. *Kona Powder and Particle Journal, 31*, 10–21.

Sayes, C. M., & Lujan, H. (2017). Characterizing the nano-bio interface using microscopic techniques: Imaging the cell system is just as important as imaging the nanoparticle system. *Current Protocols in Chemical Biology, 9*, 213–231.

Sayes, C. M., & Warheit, D. B. (2009). Characterization of nanomaterials for toxicity assessment. *Wiley Interdisciplinary Reviews: Nanomedicine and Nanobiotechnology, 1*, 660–670.

Sayes, C., Fortner, J., Guo, W., Lyon, D., Boyd, A., Ausman, K., Tao, Y., Sitharaman, B., Wilson, L., Hughes, J., West, J., & Colvin, V. (2004). The differential cytotoxicity of water-soluble fullerenes. *Nano Letters, 4*, 1881–1887.

Sayes, C., Liang, F., Hudson, J., Mendez, J., Guo, W., Beach, J., Moore, V., Doyle, C., West, J., Billups, W., Ausman, K., & Colvin, V. (2006). Functionalization density dependence of single-walled carbon nanotubes cytotoxicity in vitro. *Toxicology Letters, 161*, 135–142.

Sayes, C. M., Marchione, A. A., Reed, K. L., & Warheit, D. B. (2007a). Comparative pulmonary toxicity assessments of C60 water suspensions in rats: Few differences in fullerene toxicity in vivo in contrast to in vitro profiles. *Nano Letters, 7*, 2399–2406.

Sayes, C. M., Reed, K. L., & Warheit, D. B. (2007b). Assessing toxicity of fine and nanoparticles: Comparing in vitro measurements to in vivo pulmonary toxicity profiles. *Toxicological Sciences, 97*, 163–180.

Sharma, V., Siskova, K., Zboril, R., & Gardea-Torresdey, J. (2014). Organic-coated silver nanoparticles in biological and environmental conditions: Fate, stability and toxicity. *Advances in Colloid and Interface Science, 204*, 15–34.

Sharma, V., Yang, X., Cizmas, L., McDonald, T., Luque, R., Sayes, C., Yuan, B., & Dionysiou, D. (2017). Impact of metal ions, metal oxides, and nanoparticles on the formation of disinfection by-products during chlorination. *Chemical Engineering Journal, 317*, 777–792.

Shenoy, D., Fu, W., Li, J., Crasto, C., Jones, G., Dimarzio, C., Sridhar, S., & Amiji, M. (2006). Surface functionalization of gold nanoparticles using hetero-bifunctional poly (ethylene glycol) spacer for intracellular tracking and delivery. *International Journal of Nanomedicine, 1*, 51.

Slowing, I., Trewyn, B. G., & Lin, V. S.-Y. (2006). Effect of surface functionalization of MCM-41-type mesoporous silica nanoparticles on the endocytosis by human cancer cells. *Journal of the American Chemical Society, 128*, 14792–14793.

Sperling, R. A., & Parak, W. (2010). Surface modification, functionalization and bioconjugation of colloidal inorganic nanoparticles. *Philosophical Transactions of the Royal Society of London A: Mathematical, Physical and Engineering Sciences, 368*, 1333–1383.

Storm, G., Belliot, S. O., Daemen, T., & Lasic, D. D. (1995). Surface modification of nanoparticles to oppose uptake by the mononuclear phagocyte system. *Advanced Drug Delivery Reviews, 17*, 31–48.

Topuz, E., & Van Gestel, C. A. (2015). Toxicokinetics and toxicodynamics of differently coated silver nanoparticles and silver nitrate in Enchytraeus crypticus upon aqueous exposure in an inert sand medium. *Environmental Toxicology and Chemistry, 34*, 2816–2823.

Tsai, S., Hofmann, M., Hallock, M., Ada, E., Kong, J., & Ellenbecker, M. (2009). Characterization and evaluation of nanoparticle release during the synthesis of single-walled and multiwalled carbon nanotubes by chemical vapor deposition. *Environmental Science & Technology, 43*, 6017–6023.

Tsai, C., Huang, C., Chen, S., Ho, C., Huang, C., Chen, C., Chang, C., Tsai, S., & Ellenbecker, M. (2011). Exposure assessment of nano-sized and respirable particles at different workplaces. *Journal of Nanoparticle Research, 13*, 4161–4172.

Villanueva, A., Canete, M., Roca, A. G., Calero, M., Veintemillas-Verdaguer, S., Serna, C. J., Del Puerto Morales, M., & Miranda, R. (2009). The influence of surface functionalization on the enhanced internalization of magnetic nanoparticles in cancer cells. *Nanotechnology, 20*, 115103.

Warheit, D. B. (2008). How meaningful are the results of nanotoxicity studies in the absence of adequate material characterization? *Toxicological Sciences, 101*, 183–185.

Warheit, D. B., Hoke, R. A., Finlay, C., Donner, E. M., Reed, K. L., & Sayes, C. M. (2007). Development of a base set of toxicity tests using ultrafine TiO_2 particles as a component of nanoparticle risk management. *Toxicology Letters, 171*, 99–110.

Warheit, D. B., Reed, K. L., & Sayes, C. M. (2009a). A role for nanoparticle surface reactivity in facilitating pulmonary toxicity and development of a base set of hazard assays as a component of nanoparticle risk management. *Inhalation Toxicology, 21*(Suppl 1), 61–67.

Warheit, D. B., Sayes, C. M., & Reed, K. L. (2009b). Nanoscale and fine zinc oxide particles: Can in vitro assays accurately forecast lung hazards following inhalation exposures? *Environmental Science & Technology, 43*, 7939–7945.

Wijnhoven, S., Peijnenburg, W., Herberts, C., Hagens, W., Oomen, A., Heugens, E., Roszek, B., Bisschops, J., Gosens, I., Van De Meent, D., Dekkers, S., De Jong, W., Van Zijverden, M., Sips, A., & Geertsma, R. (2009). Nano-silver—A review of available data and knowledge gaps in human and environmental risk assessment. *Nanotoxicology, 3*, 109–U78.

Zeidan, E., Kepley, C., Sayes, C., & Sandros, M. (2015). Surface plasmon resonance: A label-free tool for cellular analysis. *Nanomedicine, 10*, 1833–1846.

Zhang, F., Durham, P., Sayes, C. M., Lau, B. L., & Bruce, E. D. (2015). Particle uptake efficiency is significantly affected by type of capping agent and cell line. *Journal of Applied Toxicology, 35*, 1114–1121.

Chapter 5
Plant Virus-Based Nanotechnologies

Amy M. Wen, Karin L. Lee, and Nicole F. Steinmetz

Nanoscale engineering is revolutionizing disease detection and prevention. Viruses have made a remarkable contribution to these developments because they can function as prefabricated nanoparticles that have naturally evolved to deliver cargos to cells and tissues. The Steinmetz Lab has established a library of plant virus-based nanoparticles and carried out comprehensive structure–function studies that have shown how to tailor these nanomaterials appropriately for biomedical applications. By exploiting the benefits of synthetic and chemical biology, plant virus-based nanotechnologies are being developed for applications in molecular imaging and drug delivery, and as cancer vaccines and immunotherapies.

Viruses as Tools for Imaging and Drug Delivery

Viruses have evolved as shielded transporters for delicate nucleic acids, which are delivered to vulnerable host cells to initiate infections (Koonin et al. 2006). The same properties that allow viruses to deliver nucleic acids can also be exploited to deliver imaging reagents and drugs. From a materials perspective, viruses can be thought of as nanoparticles with sophisticated, well-defined structures that can be

A. M. Wen
Wyss Institute for Biologically Inspired Engineering, Harvard University, Boston, MA, USA

K. L. Lee
Laboratory of Tumor Immunology and Biology, Center for Cancer Research, National Cancer Institute, NIH, Bethesda, MD, USA

N. F. Steinmetz (✉)
Department of NanoEngineering, Moores Cancer Center, University of California-San Diego, La Jolla, CA, USA
e-mail: nsteinmetz@ucsd.edu

© Springer Nature Switzerland AG 2020
P. M. Norris, L. E. Friedersdorf (eds.), *Women in Nanotechnology*, Women in Engineering and Science, https://doi.org/10.1007/978-3-030-19951-7_5

used as scaffolds for the display and encapsulation of a vast array of molecules (Pokorski and Steinmetz 2011).

An important consideration when using nanoparticles as delivery vehicles is their ability to find specific target cells. The size, shape, and charge of the particles, along with modifications that shield them from the immune system and target appropriate receptors, all play a role in their interactions within the body and therefore their specificity (Wen et al. 2013). Tailoring the precise shape of nanoscale particles in a reproducible manner presents a significant challenge. The materials provided by nature can be taken advantage of to address this challenge, since virus-based nanoparticles inherently possess different surface characteristics and self-assemble into diverse shapes and sizes (Fig. 5.1). Icosahedral, rod-shaped, and filamentous viruses are the most common, but other remarkable structures such as zipper-like and bottle-shaped viruses also exist (Prangishvili and Garrett 2005; Prangishvili et al. 2006). The ideal virus-based nanoparticle depends on the ultimate application. For example, the filamentous plant virus potato virus X (PVX) has the ability to penetrate deeply into the core of tumors, whereas the icosahedral cowpea mosaic virus (CPMV) tends to accumulate peripherally (Shukla et al. 2013).

As well as using the particles provided by nature, knowledge of the genetic and biochemical principles underlying the self-assembly of viruses allows us to tune the shape and size of the particles, and to build even more complex assemblies. Rod-shaped tobacco mosaic virus (TMV) particles can undergo a shape transition under controlled higher temperature conditions to form spheres (Bruckman et al. 2014b, 2015). Using synthetic polymeric linkers, CPMV particles can be assembled to form chains from initial icosahedral subunits (Wen and Steinmetz 2014).

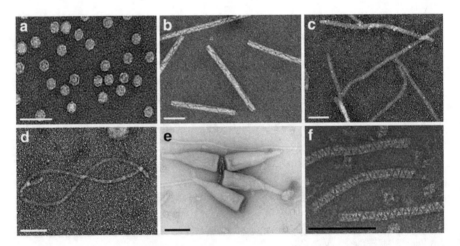

Fig. 5.1 The diversity of virus shapes and sizes. Transmission electron micrographs illustrate some of the many different virus shapes that can be found in nature. (**a**) Icosahedral cowpea mosaic virus (CPMV). (**b**) Rod-shaped tobacco mosaic virus (TMV). (**c**) Filamentous potato virus X (PVX). (**d**) Highly flexible grapevine virus A (GVA) (**a–d** reprinted with permission (Wen et al. 2013)). (**e**) *Acidianus* bottle-shaped virus (reprinted with permission (Prangishvili et al. 2006)). (**f**) *Acidianus* zipper-like virus (reprinted with permission (Rachel et al. 2002)). Scale bars = 100 nm

Additionally, TMV rods of different lengths can be formed by changing the size of the RNA molecule around which they assemble (Shukla et al. 2015). Indeed, RNA-directed self-assembly can also be used to form more exquisite structures, such as TMV-based nano-boomerangs and tetrapods (Eber et al. 2014).

After considering the desired features of the nanoparticles, the next step is to augment them with an imaging agent or drug as a new cargo. Viruses lend themselves to multiple modification strategies, including biological, chemical, and physical methods (Fig. 5.2). Virus coat proteins are encoded by the genomic nucleic acid sequence, which can be genetically altered to introduce or remove amino acids that confer certain functionalities (Miller et al. 2007; Geiger et al. 2013). For example, cysteine residues can be conjugated using thiol-based chemistries, and they also bind naturally to gold surfaces (Wang et al. 2002; Klem et al. 2003). Other amino acids are compatible with different chemistries, and this means that in many cases the virus capsid has accessible internal and external surfaces that can be modified

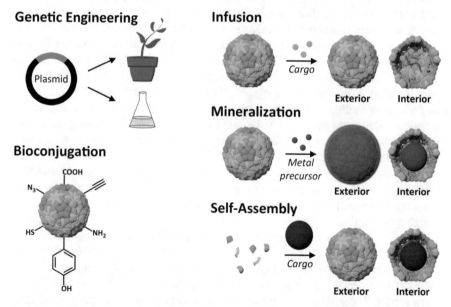

Fig. 5.2 Strategies for the modification of virus nanoparticles. Viral capsids can be modified by *genetic engineering* to insert and/or delete amino acid sequences of various functionalities. These native or genetically engineered functional groups are displayed on the capsid, allowing the *bioconjugation* of fluorescent dyes, targeting peptides, and other molecules such as drugs. Some common functionalities for bioconjugation include amines (–NH$_2$) from lysine residues, thiols (–SH) from cysteine residues, carboxylic acids (–COOH) from aspartic and glutamic acid residues, phenol rings (aromatic rings containing –OH) of tyrosine residues, as well as azides (–N$_3$) and alkynes (C–C triple bond) that can be incorporated using unnatural amino acids or through another conjugation step. Small cargo molecules can be *infused* into the intact virus and retained by electrostatic or affinity interactions. Additionally, shape-constrained *mineralization* can be achieved by exploiting the interactions between viruses and metal ions as nucleation sites. Finally, virus coat proteins can be directed to *self-assemble* around cargo molecules. Reprinted with permission (Wen and Steinmetz 2016)

separately using different methods, e.g., allowing the encapsulation of drugs and the simultaneous external display of targeting ligands (Bruckman et al. 2008; Wen et al. 2012). Compounds can also be captured within virus particles by infusion, taking advantage of affinity interactions with either the nucleic acid or the internally conjugated polymers (Yildiz et al. 2013; Wen et al. 2015a; Hovlid et al. 2014). Interactions between the capsid and metal ions can be used to introduce nucleation sites for shape-constrained mineralization (Douglas et al. 2002; Knez et al. 2003; Aljabali et al. 2011). Finally, physically disassembling the capsid and reassembling it around cargo molecules are another common modification strategy (Huang et al. 2007; Loo et al. 2007; Luque et al. 2013).

With these strategies in hand, diverse virus-based nanoparticles can be created for imaging and therapeutic applications. Fluorescence imaging is generally the simplest way to characterize nanoparticle behavior in vitro and in vivo, and new structures therefore tend to be conjugated to fluorophores for initial testing. The conjugation of virus coat proteins to fluorophores allows the particles to be tracked at the cellular and subcellular levels. Techniques such as flow cytometry can therefore be used to provide quantitative information about particle–cell interactions, and fluorescence microscopy shows in detail how the particles are taken up by cells and reveals their subsequent fate. Imaging techniques can also be used in preclinical animal models to track the fate of particles in vivo. Fluorescent virus nanoparticles have therefore been used widely, e.g., to assess changes in the circulation properties of PVX after multiple administrations (Shukla et al. 2016), and to compare the thrombus homing properties of TMV and CPMV, with TMV achieving better performance probably due to its greater capacity to drift to the side of blood vessels and interact with endothelial cells (Wen et al. 2015b).

Virus nanoparticles have also been loaded with magnetic resonance imaging (MRI) contrast agents and positron-emission tomography (PET) radiotracers, which are more relevant in the clinic because they allow the noninvasive, high-contrast imaging of soft tissues. Examples include the encapsulation of paramagnetic gadolinium (Gd) for enhanced brightness in T_1-weighted MRI, and superparamagnetic iron oxide for decreased brightness in T_2-weighted MRI. In one study, TMV was externally modified with a targeting peptide and internally conjugated with Gd(DOTA), a chelated form of Gd that can be packed tightly inside the cavity of a virus capsid. These particles were able to delineate regions containing atherosclerotic plaques at concentrations 400 times lower than clinical angiography doses (Fig. 5.3) (Bruckman et al. 2014a). Similarly, bacteriophage MS2 capsids modified with [64]Cu(DOTA) have been used to determine biodistribution of the particles in initial PET imaging studies (Farkas et al. 2013).

Engineering strategies similar to those used for imaging reagents have been applied to achieve the delivery of drugs, particularly in the fields of photodynamic therapy (PDT) and chemotherapy. PDT involves the delivery of an inert drug which is activated by light. Recent studies have shown that CPMV (Wen et al. 2016) and TMV (Lee et al. 2016) can be loaded with zinc porphyrin photosensitizers that are subsequently activated by white light and generate localized reactive oxygen species (ROS) for the destruction of melanoma cells. In the field of chemotherapy, initial studies demonstrating the efficacy of plant virus nanoparticles in preclinical

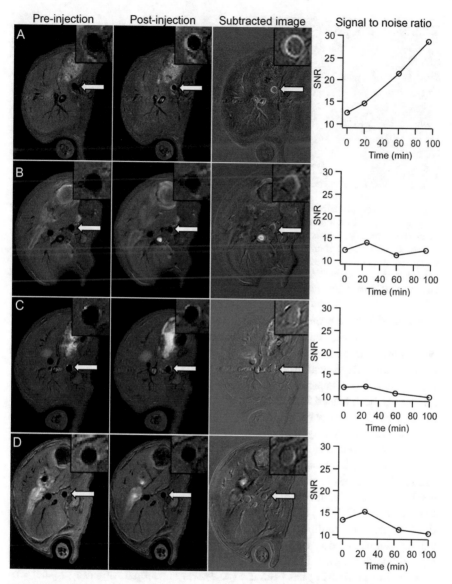

Fig. 5.3 Magnetic resonance imaging of atherosclerotic plaques. MRI was conducted before and after the injection of (**a**) targeted TMV with Gd(DOTA) contrast agent, (**b**) Gd(DOTA) alone, and (**c**) PBS in an apoE$^{-/-}$ mouse model of atherosclerosis. Insets show a magnified view of the abdominal aorta, and an enhanced signal-to-noise ratio is observed for Gd(DOTA) delivered with TMV (reprinted with permission (Bruckman et al. 2014a))

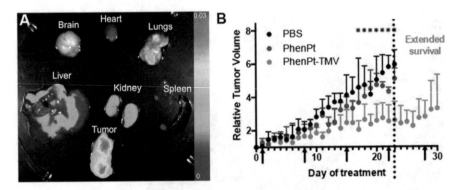

Fig. 5.4 Therapeutic delivery of phenanthriplatin in a mouse breast cancer model. (**a**) Ex vivo imaging of organs 24 h after the administration of Cy5 dye-labeled TMV loaded with phenanthriplatin (PhenPt-TMV) reveals the accumulation of particles in the tumor tissue. (**b**) Tumor volumes over time, with treatment on the days indicated by arrows. Volume was normalized to the initial tumor size at the time of treatment. Compared to phenanthriplatin (PhenPt) alone, mice treated with PhenPt-TMV showed slower tumor growth and survived longer (*$p < 0.05$). Reprinted with permission (Czapar et al. 2016)

mouse models have been published in the last couple of years. In a mouse model of triple-negative breast cancer, the novel and potent platinum(II) anticancer drug phenanthriplatin was conjugated to TMV and achieved better accumulation in the tumor tissue than phenanthriplatin alone, resulting in a fourfold reduction in tumor growth (Fig. 5.4) (Czapar et al. 2016). Likewise, the common anticancer drug doxorubicin has been loaded into PVX nanoparticles, which reduced tumor growth in a mouse xenograft model of breast cancer (Le et al. 2017).

As well as their ability to treat diseases in animals, plant virus nanoparticles have also been recruited to tackle diseases in plants. Parasitic nematodes affect crop production worldwide and are a huge burden in the agricultural industry, but the application of pesticides to the soil is inefficient because the nematodes are protected inside the plant (Quentin et al. 2013). To tackle this challenge, red clover necrotic mosaic virus (RCNMV) has been developed as a nanocarrier for the biological pesticide abamectin, resulting in healthier root growth and a lower incidence of root galling compared to untreated tomato seedlings (Cao et al. 2015). In a similar approach with the potential for rapid translation to the market, tobacco mild green mosaic virus (TMGMV) has been used to deliver the antiparasitic drug crystal violet, achieving greater bioavailability, treatment efficacy, and soil mobility compared to treatment with crystal violet alone (Chariou and Steinmetz 2017).

Viruses as Vaccines and Immunotherapies

For centuries, viruses have been used for the development of vaccines, starting with Edward Jenner, who noticed that milkmaids infected with cowpox virus were less susceptible to the more virulent smallpox virus (Riedel 2005). Since then, viruses have been developed as vaccines to combat a variety of infectious diseases, including polio, measles, mumps, rubella, influenza, human immunodeficiency virus/acquired immunodeficiency syndrome (HIV/AIDS), and Ebola hemorrhagic fever (Knobler et al. 2002; Adis International Ltd 2003; Henao-Restrepo et al. 2017; López-Macías et al. 2011). More recently, the immunostimulatory nature of viruses has also been exploited to treat noninfectious diseases, such as cancer.

Viruses have properties that make them ideal for the stimulation of the immune system. They are taken up efficiently by antigen-presenting cells due to their size, often leading to cross talk and activation of the adaptive immune system. The highly ordered, multivalent proteinaceous structures are recognized as danger signals, which can stimulate strong cellular and humoral immune responses (Lua et al. 2014). Finally, the nucleic acid contained within the protein shell is also a strong immunostimulatory agent (a pathogen-associated molecular pattern) and is a natural ligand for Toll-like receptors (Hou et al. 2011; Heil et al. 2004; Jegerlehner et al. 2007).

Building on their immunostimulatory properties, plant viruses and bacteriophages have been developed as display and delivery platforms for disease-specific antigens in a repetitive array, thus inducing effective B-cell activation via cross-linking and also epitope-specific priming T-cell responses. For example, clinically approved prophylactic vaccines against human papillomavirus (HPV), which causes cervical cancer (Harper 2009), and hepatitis B virus (HBV), which causes liver cancer (Gerlich 2015), have been developed using virus-like particles (VLPs) displaying immunogenic component proteins. Although these preventative vaccines have reduced the incidence of the corresponding cancers, there is still a need for therapeutic vaccines that can be applied when the disease is already established. Toward this goal, plant viruses have been engineered to display cancer-specific epitopes. The multivalent display helps to break self-tolerance and induce the immune system to activate epitope-specific humoral and/or cellular responses. In one example, plant viruses have been developed into cancer vaccines targeting HER2$^+$ tumors. These are conventionally treated with the clinically approved monoclonal antibodies trastuzumab and pertuzumab, but the antibodies must be administered regularly over long treatment cycles and they induce neither cellular immune responses nor the immune memory that could protect patients from metastasis and recurrence (Jhaveri and Esteva 2014; Hortobagyi 2005). A therapeutic vaccine for HER2$^+$ cancer would remove the need for repeated administration and would stimulate both cellular and humoral immune responses. Toward this long-term goal, icosahedral CPMV and filamentous PVX particles were tested to see which would be the more efficient candidate to present and deliver HER2-derived epitopes to trigger humoral responses against HER2$^+$ tumors. Structure–function studies revealed that CPMV was more efficient at presenting B-cell HER-2 epitopes, and the differences in efficacy were attributed to the distinct shapes and molecular compositions of the nanoparticles, which in turn affected their in vivo distribution, cellular uptake, and immunogenicity (Shukla et al. 2017). Building upon this knowledge, translational development has been initiated with efficacy testing in mouse HER2$^+$ cancer models. In addition to cancer, viruses have been developed to display epitopes related to chronic diseases, including hypertension (Tissot et al. 2008), Alzheimer's disease

(Chackerian et al. 2006), rheumatoid arthritis (Spohn et al. 2008), myocarditis (Sonderegger et al. 2006), and obesity (Fulurija et al. 2008). The potential applications of this approach appear limitless.

A remarkable new application of plant virus-based nanotechnology in cancer immunotherapy is based on the recent discovery that plant VLPs stimulate a potent antitumor immune response when applied as in situ vaccines, i.e., a vaccine injected directly into the tumor microenvironment, changing the environment from immunosuppressive to immunostimulatory by exploiting the tumor as a source of antigens. This modulates the local microenvironment to relieve tumor-generated immunosuppression and potentiate antitumor immunity against antigens expressed by the tumor. It is important to recognize that the antitumor response is *not* limited to the treatment of the identified, injectable tumor. Indeed, data indicate that CPMV particles induce a systemic, immune-mediated antitumor response against unrecognized metastases and protect patients from recurrence (Fig. 5.5) (Lizotte et al.

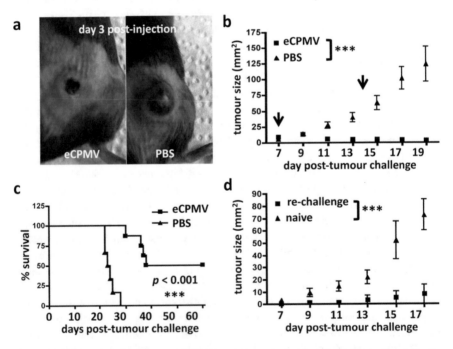

Fig. 5.5 eCPMV (=empty, RNA-free CPMV) induces systemic, durable antitumor immunity. (**a, b**) Mice bearing intradermal flank B16F10 tumors directly injected with eCPMV (arrows indicate treatment days) showed noticeably delayed tumor progression relative to PBS-injected controls ($n = 8$ eCPMV, 6 PBS). (**c**) Half of eCPMV-treated mice experienced complete elimination of primary tumors ($n = 8$ eCPMV, 6 PBS). (**d**) The majority of mice cured of primary tumors by eCPMV treatment and rechallenged on the opposite flank 4 weeks later failed to develop new tumors ($n = 4$/group). Survival analysis was carried out using the log-rank Mantel-Cox test and flank tumor growth curves were analyzed using two-way analysis of variance (*$p < 0.05$, **$p < 0.01$, and ***$p < 0.001$). Reproduced with permission (Lizotte et al. 2015)

2015). Potent efficacy has been demonstrated in several mouse models, including models of breast cancer, ovarian cancer, melanoma, and colon cancer.

The potential of viruses in cancer immunotherapy has already been recognized using mammalian vectors; hence talimogene laherparepvec (T-VEC) was approved as an in situ vaccine for non-resectable melanoma. This oncolytic viral therapy is based on herpes simplex virus that expresses granulocyte-macrophage colony-stimulating factor (GM-CSF) (Kohlhapp et al. 2015). Nevertheless, there may be advantages to replacing the mammalian virus with a plant virus. The latter can be produced by molecular farming in plants, which achieves higher yields than mammalian cell culture and does not require expensive cell culture media. More importantly, all plant viruses are noninfectious in humans, offering another layer of safety that cannot be matched by animal viruses. Even so, no therapeutic vaccines based on a plant virus have yet reached the clinic.

Although unmodified viruses have shown success as vaccines and immunotherapies, their efficacy can be improved by combining them with other immunotherapies and chemotherapies. For example, in situ vaccination using papaya mosaic virus (PapMV) was more potent when combined with either a dendritic cell-based vaccine or PD-1 blockade (Lebel et al. 2016). Other combinations include the use of plant virus particles, here PVX, and drugs such as doxorubicin. Importantly, co-administered PVX+DOX was better for in situ vaccination than PVX loaded with DOX (PVX-DOX), i.e., synergy and therefore higher efficacy were apparent when the two treatment regimens were applied separately but concurrently, but not when the drug was encapsulated in the particles. While the nanomedicine field strives to design multifunctional nanoparticles that integrate several functions and therapeutic regimens into single nanoparticle, data suggest a paradigm shift—some therapeutics may need to be administered separately to achieve the most potent therapeutic outcome (Lee et al. 2017).

Outlook

Plant virus nanoparticles provide an intriguing platform technology, which has been enhanced by comprehensive structure–function studies from which design rules have emerged to define the most suitable viruses for different applications. When targeting the vasculature for molecular imaging, the rigid nanotubes formed by TMV show advantageous vessel margination properties while avoiding immune clearance. In terms of drug delivery, the filamentous structures of PVX achieve enhanced tissue transport. Finally, icosahedral particles such as CPMV are immunostimulatory and are therefore promising candidates for immunotherapies and vaccines.

Nevertheless, the surface of plant virus nanotechnology has only been scratched and many further discoveries await. Plant viruses combine the advantages of viral and nonviral delivery systems because they do not infect or replicate in mammals, and they do not integrate into the genome, thereby offering improved safety. These protein-based carriers show exceptional stability in biological media, and shape-engineering and tailor-made stealth coatings reduce their immunogenicity. Scalable manufacturing can be achieved by molecular farming in plants, and the genetically encoded capsids offer a high degree of quality control and quality assurance because the precise structure is inherently reproducible. The potential of plant virus nanotechnology has been highlighted in several proof-of-concept and applied research studies. It is time to focus on critical toxicology and pharmacology studies beyond small animal models with the goal to translate these promising products and applications into the clinic and field-based applications.

Acknowledgements This was funded in part by the following grants to N.F.S.: NIH R01-CA224605, NIH R01-HL137674, NIH U01-CA218292, NIH R21-EB024874, NIH R01-CA202814, NIH R21-HL121130, NIH R21-EB020946, American Cancer Society 128319-RSG-15-144-01-CDD, NSF DMR-1452257 CAREER, NSF CMMI-1333651, NSF CHE-1306447, and Susan G. Komen CCR14298962. A.M.W. was supported through the following fellowships: NIH T32 EB007509, AHA 15PRE25710044, and NIH F31 HL129703, and K.L.L. was supported by the following fellowships: NIH T32 EB007509 and NIH R25 CA148052.

References

Adis International Ltd. (2003). HIV Gp120 vaccine - VaxGen: AIDSVAX, AIDSVAX B/B, AIDSVAX B/E, HIV Gp120 Vaccine - Genentech, HIV Gp120 Vaccine AIDSVAX - VaxGen, HIV Vaccine AIDSVAX – VaxGen. *Drugs in R&D, 4*(4), 249–253.

Aljabali, A. A. A., Lomonossoff, G. P., & Evans, D. J. (2011). CPMV-polyelectrolyte-templated gold nanoparticles. *Biomacromolecules, 12*(7), 2723–2728.

Bruckman, M. A., Kaur, G., Lee, L. A., Xie, F., Sepulveda, J., Breitenkamp, R., Zhang, X., Joralemon, M., Russell, T. P., Emrick, T., & Wang, Q. (2008). Surface modification of tobacco mosaic virus with "click" chemistry. *Chembiochem: A European Journal of Chemical Biology, 9*(4), 519–523.

Bruckman, M. A., Jiang, K., Simpson, E. J., Randolph, L. N., Luyt, L. G., Yu, X., & Steinmetz, N. F. (2014a). Dual-modal magnetic resonance and fluorescence imaging of atherosclerotic plaques in vivo using VCAM-1 targeted tobacco mosaic virus. *Nano Letters, 14*(3), 1551–1558.

Bruckman, M. A., Randolph, L. N., VanMeter, A., Hern, S., Shoffstall, A. J., Taurog, R. E., & Steinmetz, N. F. (2014b). Biodistribution, pharmacokinetics, and blood compatibility of native and PEGylated tobacco mosaic virus nano-rods and -spheres in mice. *Virology, 449*, 163–173.

Bruckman, M. A., VanMeter, A., & Steinmetz, N. F. (2015). Nanomanufacturing of tobacco mosaic virus-based spherical biomaterials using a continuous flow method. *ACS Biomaterials Science & Engineering, 1*(1), 13–18.

Cao, J., Guenther, R. H., Sit, T. L., Lommel, S. A., Opperman, C. H., & Willoughby, J. A. (2015). Development of abamectin loaded plant virus nanoparticles for efficacious plant parasitic nematode control. *ACS Applied Materials & Interfaces, 7*(18), 9546–9553.

Chackerian, B., Rangel, M., Hunter, Z., & Peabody, D. S. (2006). Virus and virus-like particle-based immunogens for Alzheimer's disease induce antibody responses against amyloid-beta without concomitant T cell responses. *Vaccine, 24*(37–39), 6321–6331.

Chariou, P. L., & Steinmetz, N. F. (2017). Delivery of pesticides to plant parasitic nematodes using tobacco mild green mosaic virus as a nanocarrier. *ACS Nano, 11*(5), 4719–4730.

Czapar, A. E., Zheng, Y.-R., Riddell, I. A., Shukla, S., Awuah, S. G., Lippard, S. J., & Steinmetz, N. F. (2016). Tobacco mosaic virus delivery of phenanthriplatin for cancer therapy. *ACS Nano, 10*(4), 4119–4126.

Douglas, T., Strable, E., Willits, D., Aitouchen, A., Libera, M., & Young, M. (2002). Protein engineering of a viral cage for constrained nanomaterials synthesis. *Advanced Materials, 14*(6), 415–418.

Eber, F. J., Eiben, S., Jeske, H., & Wege, C. (2014). RNA-controlled assembly of tobacco mosaic virus-derived complex structures: From nanoboomerangs to tetrapods. *Nanoscale, 7*(1), 344–355.

Farkas, M. E., Aanei, I. L., Behrens, C. R., Tong, G. J., Murphy, S. T., O'Neil, J. P., & Francis, M. B. (2013). PET imaging and biodistribution of chemically modified bacteriophage MS2. *Molecular Pharmaceutics, 10*(1), 69–76.

Fulurija, A., Lutz, T. A., Sladko, K., Osto, M., Wielinga, P. Y., Bachmann, M. F., & Saudan, P. (2008). Vaccination against GIP for the treatment of obesity. *PloS One, 3*(9), e3163.

Geiger, F. C., Eber, F. J., Eiben, S., Mueller, A., Jeske, H., Spatz, J. P., & Wege, C. (2013). TMV nanorods with programmed longitudinal domains of differently addressable coat proteins. *Nanoscale, 5*(9), 3808–3816.

Gerlich, W. H. (2015). Prophylactic vaccination against hepatitis B: Achievements, challenges and perspectives. *Medical Microbiology and Immunology, 204*(1), 39–55.

Harper, D. M. (2009). Currently approved prophylactic HPV vaccines. *Expert Review of Vaccines, 8*(12), 1663–1679.

Heil, F., Hemmi, H., Hochrein, H., Ampenberger, F., Kirschning, C., Akira, S., Lipford, G., Wagner, H., & Bauer, S. (2004). Species-specific recognition of single-stranded RNA via toll-like receptor 7 and 8. *Science (New York, N.Y.), 303*(5663), 1526–1529.

Henao-Restrepo, A. M., Camacho, A., Longini, I. M., Watson, C. H., Edmunds, W. J., Egger, M., Carroll, M. W., Dean, N. E., Diatta, I., Doumbia, M., Draguez, B., Duraffour, S., Enwere, G., Grais, R., Gunther, S., Gsell, P.-S., Hossmann, S., Watle, S. V., Kondé, M. K., Kéïta, S., Kone, S., Kuisma, E., Levine, M. M., Mandal, S., Mauget, T., Norheim, G., Riveros, X., Soumah, A., Trelle, S., Vicari, A. S., Røttingen, J.-A., & Kieny, M.-P. (2017). Efficacy and effectiveness of an RVSV-vectored vaccine in preventing Ebola virus disease: Final results from the Guinea ring vaccination, open-label, cluster-randomised trial (Ebola Ça Suffit!). *The Lancet, 389*(10068), 505–518.

Hortobagyi, G. N. (2005). Trastuzumab in the treatment of breast cancer. *New England Journal of Medicine, 353*(16), 1734–1736.

Hou, B., Saudan, P., Ott, G., Wheeler, M. L., Ji, M., Kuzmich, L., Lee, L. M., Coffman, R. L., Bachmann, M. F., & DeFranco, A. L. (2011). Selective utilization of toll-like receptor and MyD88 signaling in B cells for enhancement of the antiviral germinal center response. *Immunity, 34*(3), 375–384.

Hovlid, M. L., Lau, J. L., Breitenkamp, K., Higginson, C. J., Laufer, B., Manchester, M., & Finn, M. G. (2014). Encapsidated atom-transfer radical polymerization in Qβ virus-like nanoparticles. *ACS Nano, 8*(8), 8003–8014.

Huang, X., Bronstein, L. M., Retrum, J., Dufort, C., Tsvetkova, I., Aniagyei, S., Stein, B., Stucky, G., McKenna, B., Remmes, N., Baxter, D., Kao, C. C., & Dragnea, B. (2007). Self-assembled virus-like particles with magnetic cores. *Nano Letters, 7*(8), 2407–2416.

Jegerlehner, A., Maurer, P., Bessa, J., Hinton, H. J., Kopf, M., & Bachmann, M. F. (2007). TLR9 signaling in B cells determines class switch recombination to IgG2a. *Journal of Immunology (Baltimore, MD: 1950), 178*(4), 2415–2420.

Jhaveri, K., & Esteva, F. J. (2014). Pertuzumab in the treatment of HER2+ breast cancer. *Journal of the National Comprehensive Cancer Network: JNCCN, 12*(4), 591–598.

Klem, M. T., Willits, D., Young, M., & Douglas, T. (2003). 2-D array formation of genetically engineered viral cages on Au surfaces and imaging by atomic force microscopy. *Journal of the American Chemical Society, 125*(36), 10806–10807.

Knez, M., Bittner, A. M., Boes, F., Wege, C., Jeske, H., Maiβ, E., & Kern, K. (2003). Biotemplate synthesis of 3-Nm nickel and cobalt nanowires. *Nano Letters, 3*(8), 1079–1082.

Knobler, S., Lederberg, J., Pray, L. A., & Institute of Medicine (U.S.) (Eds.). (2002). *Considerations for viral disease eradication: Lessons learned and future strategies: Workshop summary.* Washington, DC: National Academy Press.

Kohlhapp, F. J., Zloza, A., & Kaufman, H. L. (2015). Talimogene laherparepvec (T-VEC) as cancer immunotherapy. *Drugs of Today (Barcelona, Spain: 1998), 51*(9), 549–558.

Koonin, E. V., Senkevich, T. G., & Dolja, V. V. (2006). The ancient virus world and evolution of cells. *Biology Direct, 1*, 29.

Le, D. H. T., Lee, K. L., Shukla, S., Commandeur, U., & Steinmetz, N. F. (2017). Potato virus X, a filamentous plant viral nanoparticle for doxorubicin delivery in cancer therapy. *Nanoscale, 9*(6), 2348–2357.

Lebel, M.-È., Chartrand, K., Tarrab, E., Savard, P., Leclerc, D., & Lamarre, A. (2016). Potentiating cancer immunotherapy using papaya mosaic virus-derived nanoparticles. *Nano Letters, 16*(3), 1826–1832.

Lee, K. L., Carpenter, B. L., Wen, A. M., Ghiladi, R. A., & Steinmetz, N. F. (2016). High aspect ratio nanotubes formed by tobacco mosaic virus for delivery of photodynamic agents targeting melanoma. *ACS Biomaterials Science & Engineering, 2*(5), 838–844.

Lee, K. L., Murray, A. A., Le, D. H. T., Sheen, M. R., Shukla, S., Commandeur, U., Fiering, S., & Steinmetz, N. F. (2017). Combination of plant virus nanoparticle-based in situ vaccination with chemotherapy potentiates antitumor response. *Nano Letters, 17*(7), 4019–4028.

Lizotte, P. H., Wen, A. M., Sheen, M. R., Fields, J., Rojanasopondist, P., Steinmetz, N. F., & Fiering, S. (2015). In situ vaccination with cowpea mosaic virus nanoparticles suppresses metastatic cancer. *Nature Nanotechnology, 11*(3), 295–303.

Loo, L., Guenther, R. H., Lommel, S. A., & Franzen, S. (2007). Encapsidation of nanoparticles by red clover necrotic mosaic virus. *Journal of the American Chemical Society, 129*(36), 11111–11117.

López-Macías, C., Ferat-Osorio, E., Tenorio-Calvo, A., Isibasi, A., Talavera, J., Arteaga-Ruiz, O., Arriaga-Pizano, L., Hickman, S. P., Allende, M., Lenhard, K., Pincus, S., Connolly, K., Raghunandan, R., Smith, G., & Glenn, G. (2011). Safety and immunogenicity of a virus-like particle pandemic influenza A (H1N1) 2009 vaccine in a blinded, randomized, placebo-controlled trial of adults in Mexico. *Vaccine, 29*(44), 7826–7834.

Lua, L. H. L., Connors, N. K., Sainsbury, F., Chuan, Y. P., Wibowo, N., & Middelberg, A. P. J. (2014). Bioengineering virus-like particles as vaccines: Virus-like particles as vaccines. *Biotechnology and Bioengineering, 111*(3), 425–440.

Luque, D., de la Escosura, A., Snijder, J., Brasch, M., Burnley, R. J., Koay, M. S. T., Carrascosa, J. L., Wuite, G. J. L., Roos, W. H., Heck, A. J. R., Cornelissen, J. J. L. M., Torres, T., & Castón, J. R. (2013). Self-assembly and characterization of small and monodisperse dye nanospheres in a protein cage. *Chemical Science, 5*(2), 575–581.

Miller, R. A., Presley, A. D., & Francis, M. B. (2007). Self-assembling light-harvesting systems from synthetically modified tobacco mosaic virus coat proteins. *Journal of the American Chemical Society, 129*(11), 3104–3109.

Pokorski, J. K., & Steinmetz, N. F. (2011). The art of engineering viral nanoparticles. *Molecular Pharmaceutics, 8*(1), 29–43.

Prangishvili, D., & Garrett, R. A. (2005). Viruses of hyperthermophilic crenarchaea. *Trends in Microbiology, 13*(11), 535–542.

Prangishvili, D., Forterre, P., & Garrett, R. A. (2006). Viruses of the Archaea: A unifying view. *Nature Reviews Microbiology, 4*(11), 837–848.

Quentin, M., Abad, P., & Favery, B. (2013). Plant parasitic nematode effectors target host defense and nuclear functions to establish feeding cells. *Frontiers in Plant Science, 4*, 53.

Rachel, R., Bettstetter, M., Hedlund, B. P., Häring, M., Kessler, A., Stetter, K. O., & Prangishvili, D. (2002). Remarkable morphological diversity of viruses and virus-like particles in hot terrestrial environments. *Archives of Virology, 147*(12), 2419–2429.

Riedel, S. (2005). Edward Jenner and the history of smallpox and vaccination. *Proceedings (Baylor University Medical Center), 18*(1), 21–25.

Shukla, S., Ablack, A. L., Wen, A. M., Lee, K. L., Lewis, J. D., & Steinmetz, N. F. (2013). Increased tumor homing and tissue penetration of the filamentous plant viral nanoparticle potato virus X. *Molecular Pharmaceutics, 10*(1), 33–42.

Shukla, S., Eber, F. J., Nagarajan, A. S., DiFranco, N. A., Schmidt, N., Wen, A. M., Eiben, S., Twyman, R. M., Wege, C., & Steinmetz, N. F. (2015). The impact of aspect ratio on the bio-distribution and tumor homing of rigid soft-matter nanorods. *Advanced Healthcare Materials, 4*(6), 874–882.

Shukla, S., Dorand, R. D., Myers, J. T., Woods, S. E., Gulati, N. M., Stewart, P. L., Commandeur, U., Huang, A. Y., & Steinmetz, N. F. (2016). Multiple administrations of viral nanoparticles alter in vivo behavior—Insights from intravital microscopy. *ACS Biomaterials Science & Engineering, 2*(5), 829–837.

Shukla, S., Myers, J. T., Woods, S. E., Gong, X., Czapar, A. E., Commandeur, U., Huang, A. Y., Levine, A. D., & Steinmetz, N. F. (2017). Plant viral nanoparticles-based HER2 vaccine: Immune response influenced by differential transport, localization and cellular interactions of particulate carriers. *Biomaterials, 121*, 15–27.

Sonderegger, I., Röhn, T. A., Kurrer, M. O., Iezzi, G., Zou, Y., Kastelein, R. A., Bachmann, M. F., & Kopf, M. (2006). Neutralization of IL-17 by active vaccination inhibits IL-23-dependent autoimmune myocarditis. *European Journal of Immunology, 36*(11), 2849–2856.

Spohn, G., Keller, I., Beck, M., Grest, P., Jennings, G. T., & Bachmann, M. F. (2008). Active immunization with IL-1 displayed on virus-like particles protects from autoimmune arthritis. *European Journal of Immunology, 38*(3), 877–887.

Tissot, A. C., Maurer, P., Nussberger, J., Sabat, R., Pfister, T., Ignatenko, S., Volk, H.-D., Stocker, H., Müller, P., Jennings, G. T., Wagner, F., & Bachmann, M. F. (2008). Effect of immunisation against angiotensin II with CYT006-AngQb on ambulatory blood pressure: A double-blind, randomised, placebo-controlled phase IIa study. *The Lancet, 371*(9615), 821–827.

Wang, Q., Lin, T., Johnson, J. E., & Finn, M. G. (2002). Natural supramolecular building blocks. *Chemistry & Biology, 9*(7), 813–819.

Wen, A. M., & Steinmetz, N. F. (2014). The aspect ratio of nanoparticle assemblies and the spatial arrangement of ligands can be optimized to enhance the targeting of cancer cells. *Advanced Healthcare Materials, 3*(11), 1739–1744.

Wen, A. M., & Steinmetz, N. F. (2016). Design of virus-based nanomaterials for medicine, bio-technology, and energy. *Chemical Society Reviews, 45*(15), 4074–4126.

Wen, A. M., Shukla, S., Saxena, P., Aljabali, A. A. A., Yildiz, I., Dey, S., Mealy, J. E., Yang, A. C., Evans, D. J., Lomonossoff, G. P., & Steinmetz, N. F. (2012). Interior engineering of a viral nanoparticle and its tumor homing properties. *Biomacromolecules, 13*(12), 3990–4001.

Wen, A. M., Rambhia, P. H., French, R. H., & Steinmetz, N. F. (2013). Design rules for nanomedi-cal engineering: From physical virology to the applications of virus-based materials in medi-cine. *Journal of Biological Physics, 39*(2), 301–325.

Wen, A. M., Le, N., Zhou, X., Steinmetz, N. F., & Popkin, D. L. (2015a). Tropism of CPMV to professional antigen presenting cells enables a platform to eliminate chronic infections. *ACS Biomaterials Science & Engineering, 1*(11), 1050–1054.

Wen, A. M., Wang, Y., Jiang, K., Hsu, G. C., Gao, H., Lee, K. L., Yang, A. C., Yu, X., Simon, D. I., & Steinmetz, N. F. (2015b). Shaping bio-inspired nanotechnologies to target thrombosis for dual optical-magnetic resonance imaging. *Journal of Materials Chemistry. B, Materials for Biology and Medicine, 3*(29), 6037–6045.

Wen, A. M., Lee, K. L., Cao, P., Pangilinan, K., Carpenter, B. L., Lam, P., Veliz, F. A., Ghiladi, R. A., Advincula, R. C., & Steinmetz, N. F. (2016). Utilizing viral nanoparticle/dendron hybrid conjugates in photodynamic therapy for dual delivery to macrophages and cancer cells. *Bioconjugate Chemistry, 27*(5), 1227–1235.

Yildiz, I., Lee, K. L., Chen, K., Shukla, S., & Steinmetz, N. F. (2013). Infusion of imaging and therapeutic molecules into the plant virus-based carrier cowpea mosaic virus: Cargo-loading and delivery. *Journal of Controlled Release: Official Journal of the Controlled Release Society, 172*(2), 568–578.

Chapter 6
Power Generation Using Solid-State Heat Engines

Mona Zebarjadi

Introduction

> *We do not inherit the earth from our ancestors. We borrow it from our children.*
> *Native American proverb*

Imagine that you live in an infinitely large home. It does not matter how much trash you generate since you can always move to another side of your house, leaving your trash behind. This is how the public has treated earth, their common home, for centuries, dumping trash in the oceans and in the atmosphere. Today, humans face a rapidly increasing population, and a significant amount of trash accumulated from the past. This is the era in which communities realize that their imaginary infinite home is in fact finite and they realize that it is time to start cleaning the oceans, the atmosphere, and the land and live sustainably or else the human race will not survive.

Now, imagine a world in which all our energy needs are produced without polluting the environment. Imagine a world in which we run our cars, our houses, and our factories without burning fossil fuels. It is not difficult to imagine such a world, right? It is because many parts of the technology have already been developed! Today, the public can install solar cells on their rooftops to go off the grid. Electric, hybrid, and to some extent fuel cell cars have entered the transportation section. Many clean and renewable energy power plants, including geothermal power plants, wind farms, concentrated solar power, solar photovoltaic farms, and tidal barrages,

M. Zebarjadi (✉)
Electrical and Computer Engineering Department, University of Virginia,
Charlottesville, VA, USA

Materials Science and Engineering Department, University of Virginia,
Charlottesville, VA, USA
e-mail: m.zebarjadi@virginia.edu

© Springer Nature Switzerland AG 2020
P. M. Norris, L. E. Friedersdorf (eds.), *Women in Nanotechnology*, Women in Engineering and Science, https://doi.org/10.1007/978-3-030-19951-7_6

produce electricity at large scales to run factories. The current generation is witnessing a remarkable transition from coal, oil, and gas to clean, environmentally friendly and renewable energy systems.

This transition is not an easy one. There are two main obstacles. First, there are many beneficiaries of the fossil fuel industry standing in the way. Second, the energy landscape is just enormous! Think about how much trash each of us produces, how many trees are needed to clean up the air, and how challenging it is to replace all that consumption with clean resources. To give you an idea, in the United States 2.5 gallons of oil, 13.7 pounds of coal, and 234 cubic feet of natural gas are consumed per capita per day (DOE 2018). As a result, in total the United States dumps 21.5 metric tons of CO_2 into the atmosphere per capita per year.

A gallon of oil produces about 35 kWh of energy. Adding coal and gas numbers, on the average a US citizen uses about 230 kWh of energy per day. In comparison, a 1 m^2 commercial solar panel produces about 0.6–0.8 kWh per day. Therefore, each person roughly needs 330 solar panels to address all their energy needs. It is then not a surprise that despite all solar installations, the United States only produces 0.6% of its primary energy from solar. In fact, all clean renewable resources combined only add up to about 10% of the US primary energy resources (Muller 2008; MacKay 2010). As you see, the transition is not an easy one, and it is a challenge that requires a strong workforce. There is not a single solution, but rather a combination of technologies and visions that can resolve the problem.

Here is one aspect: more than half of the energy that is produced in power plants by burning different fuels is wasted as heat before customers use it. The losses are all the way from generation at the power plant, transportation, and storage to conversion at the customer end. Two questions arise. First, is it possible to develop technologies that are more efficient? Second, is it possible to convert a part of the wasted heat back into useful electricity? These are the questions that many scientists and engineers have struggled with. Many parts of the puzzle have been solved and many other parts remain to be solved. It is a large, difficult puzzle and a work in progress!

The answer to the first question, in some cases, is yes, and in other cases there are fundamental limits to the efficiency. The invention of LEDs is an example of replacing one kind of technology with another higher efficiency one. Thomas Edison's incandescent bulbs, and even their modern versions, are less than 10% efficient. In these bulbs, only a small fraction of the electricity turns into visible light, and the rest of it produces heat. LEDs however, are 90% efficient and a great alternative.

An example of fundamental limits is the limited efficiency of heat engines. A heat engine is a device that, when placed between a cold heat sink at T_C and a hot heat source at T_H, converts the heat flux into motive power. As an example, your car engine is a heat engine wherein the hot side is the gasoline-burning chamber and the cold side is the ambient. To have high efficiency, the maximum temperature difference between the hot side and the cold side is needed. However, it is not possible to design a heat engine with an efficiency higher than that of the Carnot efficiency,

$\eta_{\text{carnot}} = 1 - (T_\text{C} / T_\text{H})$. This efficiency limit is imposed by the laws of thermodynamics (Borgnakke and Sonntag 2013).

Despite this limited efficiency, heat engines are still useful in improving the state of energy production. In fact, they are the answer to the second question of how to convert a part of the wasted heat back into electricity. Many of the heat engines (e.g., steam engine, Stirling engine, combustion turbine, gasoline and diesel engines) convert heat to mechanical motion. Using a generator this mechanical motion can be converted into electricity. In many applications, solid-state heat engines, which can convert heat directly into electricity with no moving parts, are desirable. These solid-state devices usually require minimum maintenance and are useful for remote areas with limited access such as space or remote villages without grid access. Other example applications of solid-state heat engines include temperature sensing, industrial and automobile heat recovery, and harvesting solar energy. In the latter, the solar light is first converted into heat and then into electricity using a thermoelectric power generator (Chen 2011).

Finally, while it is not the focus of this chapter, it is important to note that most solid-state heat engines can run backward as a refrigerator. In the power generation mode, heat enters the engine at the hot side, a part of it is converted into electricity, and the rest is rejected to the cold side (Fig. 6.1a). In the refrigeration cycle, an electric current is applied (work) to pump heat from the cold side to the hot side (opposite to the natural direction of flux, Fig. 6.1b). This mode of operation is useful for design of cooling systems and portable refrigerators. The refrigerator is a thermal diode, and therefore can also be used as a thermal switch (Wehmeyer et al. 2017; Kim and Kaviany 2016; Adams et al. 2019). Upon reversing the direction of the current, heat is pumped from the hot side to the cold side (in the direction of natural heat flux, Fig. 6.1c). This last mode of operation is similar to a heat spreader

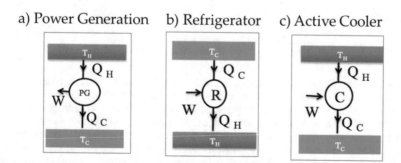

Fig. 6.1 Operational modes for heat engines. Most of these engines can operate as either power generators, refrigerators, or active coolers for efficient cooling

and is useful for fast and efficient cooling of the hot spots of a chip (electronic cooling) (Zebarjadi 2015) More applications will be discussed in the last section focused on the future of solid-state heat engines.

Types of Thermal to Electrical Power Generators with No Moving Parts

There are several types of vacuum and solid-state heat engines that convert heat directly to electricity without any moving parts. Thermoelectric power generators are the most widely used ones and therefore they will be the focus of this chapter. However, it is worth mentioning other types of solid-state power generators briefly. This section is devoted to this task.

Some power generators, such as Nernst-Ettingshausen power generators (Horst 1963) and spin Seebeck generators (Uchida et al. 2008), work under an applied magnetic field. These are not widely used. The Nernst voltage is comparable to the thermoelectric voltage (Ziabari et al. 2016), but in practice it is difficult to add a permanent magnet in the device design. Therefore, these generators only have niche applications. The spin Seebeck effect was only recently reported (Uchida et al. 2008) and the signal observed is orders of magnitude smaller compared to the thermoelectric voltage. Therefore, the spin Seebeck effect is mostly proposed for sensing applications.

There are other generators that do not require any applied magnetic fields and therefore are more practical. Figure 6.2 summarizes some of the types available. In almost all cases, these heat engines can run backward. Running backward means

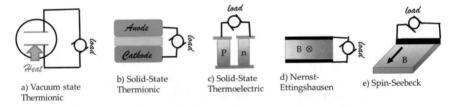

a) Vacuum state Thermionic
b) Solid-State Thermionic
c) Solid-State Thermoelectric
d) Nernst-Ettingshausen
e) Spin-Seebeck

Fig. 6.2 Different types of solid-state and vacuum-state thermal to electrical power generators (heat engines). (**a**) Vacuum-state thermionic generators, cathode (red) and anode (blue line), are made out of metals and are closely spaced inside vacuum. (**b**) Solid-state thermionic devices wherein a cathode and anode are separated by a semiconductor (pink). The semiconductor is thin (<100 nm) to ensure ballistic transport of electrons. (**c**) Solid-state thermoelectric power generator made out of p-type and n-type legs placed thermally in parallel and electrically in series. (**d**) Nernst-Ettingshausen generator: a temperature gradient ($\vec{\nabla}T$ shown by change of color from red to blue) and a perpendicular magnetic field ($\vec{B}$) are applied to a material. A voltage (current) is then generated in the direction perpendicular to both $\vec{B}$ and $\vec{\nabla}T$ fields. (**e**) Spin Seebeck generator: a temperature gradient is applied to a magnetic material and is converted to a spin voltage. An attached metallic film (black bar), usually platinum, can then transform this spin voltage into an electric voltage

one can pass current (perform work) to pump heat from the cold side to the hot side (refrigeration or cooling mode). In other words, a good heat engine (power generator) is also a good refrigerator.

Vacuum thermionic power generators (Fig. 6.2a) were first built and used as diodes by Fleming in 1905. In 1956, Hatsopoulos, a PhD student at MIT, proposed the use of these diodes as power generators in his PhD thesis (Hatsopoulos and Kaye 1958). The first demonstrations were quite successful with efficiencies as high as 13–16% (Wilson 1959; Hatsopoulos and Kaye 1958). The main problem with these devices is their very high operating temperatures. These devices work at temperatures higher than 1500 K and most metals melt at such high temperatures. For example, gold melts at 1330 K and so does copper, silver, and aluminum. Therefore, there are only limited metals available to be used as the cathode. Furthermore, most of the industrial wasted heat is at lower temperatures (<500 K) and therefore vacuum-based thermionic generators are not applicable. The reason for the high operating temperature of vacuum-based thermionic generators is the large work function of metals. Electrons need to heat up to overcome the work function and escape the surface of the cathode. Work function of most metals is well above 2 eV and therefore really high temperatures are needed as previously mentioned. To give a perspective, the kinetic energy of the electrons is proportional to their temperature and at room temperature (300 K) the average energy of the electrons is only 25 meV. Most of today's efforts are focused on designing new stable materials to be used as cathodes and anodes with lower work functions. This would enable operations at lower temperatures (Koeck et al. 2009; Massicotte et al. 2016).

The vacuum could be replaced by a solid to change these devices to solid-state thermionic power generators (SSTI, Fig. 6.2b). These generators are very attractive, since they can be fabricated at nanoscales, they do not require vacuum, and they can operate at lower temperatures. The latter is possible since the transport barrier for electrons going from a metal to a semiconductor could be designed (by choosing the proper metal-semiconductor pair) to be as small as a few meV. The main drawback of SSTIs is their low efficiency, which is the result of their large heat leak (Mahan et al. 1998). Heat can transfer much more easily in a solid (by conduction) as opposed to in vacuum (by radiation). Therefore, when vacuum is replaced by solid, the result is a much larger heat leak. One of the areas that the Energy Science, Nanotechnology and Imagination lab (E-Snail) (Zebarjadi 2019) at the University of Virginia is focused on is finding ways to minimize this heat leak (Zebarjadi 2017). One way, for example, is to use layered van der Waals structures (Wang et al. 2016). These layered structures are made out of atomically thin planes (monolayeres) which are weakly bonded and therefore are extremely poor thermal conductors. These weak bonds reduce the heat leak and can increase the efficiency significantly (Wang et al. 2018).

Thermoelectric modules are the most widely used solid-state thermal to electrical power generators (Goldsmid 2010; Ioffe 1957). They have been used in many space satellites and space probes (e.g., voyage 1 and 2) and have provided steady power operating between a radioactive material as the hot side and outer space as the cold side. These generators are made out of p- and n-type semiconductor legs.

When placed between hot and cold, carriers (both holes and electrons) diffuse from the hot side to the cold side, generating positive and negative charge accumulation at the cold side of the p and n legs, respectively. These positive and negative ends are just like the ends of a battery, and when connected to an outer circuit result in flow of current. The next section will provide more details about thermoelectric modules.

A Very Brief History of Thermoelectric Power Generators

The most well-known and widely studied solid-state heat engines are thermoelectric power generators. The thermoelectricity effect was discovered by Thomas Seebeck in 1821 (Ioffe 1958). He observed that a circuit made out of two dissimilar metals deflected a compass magnet when a part of the circuit was heated. Seebeck made an analogy to Oersted's observation[1] and stated that a temperature difference can create a magnetic field. Some of his colleagues pointed out that the correct explanation might be that the temperature difference produces an electric current, which in turn produces a magnetic field. Today it is known that the latter explanation is the correct one, but Seebeck never accepted it and, up until his last days, kept trying to convince others that the effect is a magnetic effect in nature. Seebeck tried many different materials, mostly metals, but also some minerals. Today, it is known that some of the materials that he used can convert heat to electricity with about 3% efficiency. Noting that at his time steam engines were also only about 3% efficient, and the only available option at the time for producing electricity was very weak electrostatic generators, it can clearly be seen that thermoelectric devices had a great chance to take over the market if they were well understood, but instead they were ignored (Ioffe 1958).

Perhaps the most important era for thermoelectricity was around 1950, when Abram Ioffe put the modern version of thermoelectricity theory together, introduced the figure of merit, and promoted the use of doped semiconductors. At Ioffe's time, thermoelectric modules with efficiencies around 5% were demonstrated and used in remote areas of Russia that did not have access to the grid. Thermoelectric power generators provided a good alternative and required a minimum level of maintenance (Ioffe 1957).

The next major breakthrough came in 1993 when Hicks and Dresselhaus introduced the idea of using lower dimensional materials such as quantum wells (2D) (Hicks and Dresselhaus 1993a), nanowires (1D) (Hicks and Dresselhaus 1993b), and quantum dots (0D) to improve the transport quality of thermoelectric legs. This idea will be discussed in more detail in the next section. Today the best thermoelectric materials are nanostructured materials which are built following Hicks and Dresselhous's original vision but have high efficiencies due to enhanced understanding of additional fundamental physics concepts as described below.

---

[1] Oersted observation was only a year before Seebeck's observation. In 1820, he reported that a magnetic needle is deflected by the flow of an electric current in a nearby conductor.

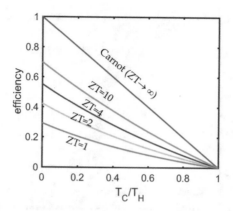

Fig. 6.3 The efficiency of an ideal thermoelectric power generator as a function of the cold-side temperature (T_C) divided by the hot-side temperature (T_H) plotted for different ZT values. Here an ideal thermoelectric power generator refers to a generator with no parasitic losses. It is assumed that the material properties are temperature independent. For a real device, this ZT could be interpreted as the average ZT of p and n legs between T_C and T_H

Physics of Thermoelectricity

A measure for the efficiency of thermoelectric modules is the dimensionless figure of merit or ZT. The efficiency of a thermoelectric power generator is a function of two parameters: the temperature ratio of the cold side to the hot side (recall the Carnot efficiency) and ZT. Figure 6.3 shows the efficiency of an ideal thermoelectric module versus the temperature ratio of the cold side to the hot side. The efficiency is plotted for different ZT values. As can be seen, efficiency is an increasing function of ZT. A figure of merit, ZT, of infinity corresponds to the Carnot efficiency. Current thermoelectric modules on the market have ZT values of about 1, while ZT values higher than 2 are reported in the laboratory (e.g., Biswas et al. 2012).

A good part of the thermoelectric research is focused on designing materials with improved ZT values. Note that ZT is defined by $ZT = \dfrac{\sigma S^2 T}{\kappa}$, where σ is electrical conductivity, S is the Seebeck coefficient, T is the temperature, and κ is the thermal conductivity. The nominator of ZT, σS^2, is called the thermoelectric power factor (PF). Electrical conductivity is a measure of how good a material conducts electricity. For example, metals have many electrons and therefore large electrical conductivity while insulators have large bandgaps and are not able to conduct electricity. The thermal conductivity is a measure of how good a material conducts heat. Materials with weak lattice bonds and heavy atoms generally have low thermal conductivity (Nolas et al. 2001).

The Seebeck coefficient is defined under open circuit conditions, and it is the measured voltage as a result of a temperature differential applied along the sample. When there is a temperature differential, carriers tend to diffuse from the hot side to the cold side. For example, in a metal there are many electrons and they form a

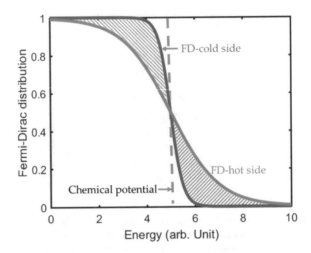

Fig. 6.4 The Fermi-Dirac (FD) distribution versus electron energy schematically plotted for the hot side and cold side of the same material. The shaded area at energies above the chemical potential times the density of state at these energies is equal to the excess number of electrons on the hot side. These electrons tend to diffuse to the cold side. Similarly, the shaded space below the chemical potential times the density of states at these energies is equal to the excess number of electrons on the cold side. These electrons tend to diffuse to the hot side. Since the two shaded spaces have exactly the same area, if the density of states is symmetric around the chemical potential, then there will be no net charge accumulations on the sides and the Seebeck voltage is zero

Fermi-Dirac distribution function. When there is a temperature difference there are more hot electrons (electrons with energies higher than the chemical potential) on the hot side compared to the cold side, resulting in the net flow of hot electrons to the cold side (see Fig. 6.4). At the same time, there are more cold electrons on the cold side compared to the hot side, and therefore these electrons flow back to the hot side. On average, if the number of hot and cold electrons is the same, there is no net voltage. The only way to have a net voltage is if there is an asymmetry between the number of hot and cold electrons. This means that the density of states (DOS) around the chemical potential should be asymmetric. The asymmetry of the DOS in semiconductors could be much larger than that in metals due to the presence of the bandgap. Therefore, semiconductors in general have Seebeck coefficient values much larger than metals. As an example, the Seebeck coefficient values of lead, copper, and platinum at room temperatures are -1.05, 1.94, and -4.92 μV/K, respectively (Rowe 2005). In contrast, the order of magnitude of the Seebeck coefficient of semiconductors can easily be as high as a hundred μV/K. Note that one cannot report a value for a specific semiconductor since the Seebeck coefficient changes significantly with the doping concentration and therefore can be adjusted in a semiconductor. Most good thermoelectric semiconductors have a Seebeck coefficient between 100 and 300 μV/K at their optimum power factor.

Lower dimensional materials have a sharper density of states. The band structure in 3D is proportional to $\sqrt{E}$, where E is the energy of the electrons. In 2D, the DOS is a series of step functions. In 1D, it is proportional to $1/\sqrt{E}$ and in 0D it is a series of delta functions (Marder 2010). These sharp features make the density of states asymmetric and therefore there is the promise of large Seebeck coefficients. This was the argument originally presented by Hicks and Dresselhaus (Hicks and Dresselhaus 1993a). They argued that the thermoelectric power factor is larger at lower dimensions due to enhanced Seebeck coefficient values. Many scientists have tried to build and measure low-dimensional materials. Despite the sufficiently small confinement lengths, only a handful of the techniques demonstrated enhanced thermoelectric power factor (Hung et al. 2016; Zebarjadi et al. 2011; Heremans et al. 2012; Curtin et al. 2013).

The discussion here will be limited to only one technique, which not only has shown great promise and has been very successfully applied to enhance the performance of thermoelectric materials, but also is cost effective. Therefore, it is the most widely used technique today in laboratories, and it has already advanced to industry. The best thermoelectric materials today are nanostructured materials which are basically made by ball milling ingots or elements to make nano-powders and then hot pressing these nano-powders into a bulk sample. The first demonstration of using this technique to improve the thermoelectric properties was in 2009. Ren, Chen, and Dresselhous groups, inspired by Hicks' and Dresselhaus' idea of lower dimensions, prepared these samples (Poudel et al. 2008). The samples are made from the state-of-the-art bismuth telluride ingots. They showed by nanostructuring that the ZT value increases by about 40%. However, upon examination of the data, they realized that the enhancement was not coming from an improved Seebeck coefficient, but rather resulted from reduced thermal conductivity. Instead of behaving as a 0D sample, this sample really behaves like a 3D sample with lots of scattering centers (interfaces or grain boundaries), which scatter phonons effectively and therefore lower the thermal conductivity.

In semiconductors, while charge carriers are electrons, heat carriers are mainly phonons. Thermoelectric semiconductors are usually heavily doped, which means that electrons participating in transport have very small mean free paths (few nanometers), and at the same time they are in a very narrow energy window (few k_BT around the chemical potential). On the contrary, phonons are bosons and they are not confined to move freely at a specific energy window. In other words, phonons of a wide energy range are free to move. As a result, phonons have a wide range of mean free paths. There is a large portion of the phonons with mean free paths larger than a few tens of nanometers (Zebarjadi et al. 2012). Therefore, structures (grain boundaries) with spacings of about a few tens of nanometers made inside a bulk sample do not affect the mean free path of electrons. However, all phonons with mean free paths larger than the structure spacing thermalize at these interfaces, and they cannot fully deliver the heat that they are carrying. Thus, proper nanostructuring does not significantly affect the electrical conductivity, but it is very effective in lowering the thermal conductivity. The figure of merit increases when the ratio of electrical conductivity to thermal conductivity increases.

After the original demonstration, nanostructuring has been applied to many different samples and has proven successful in enhancing the thermoelectric figure of merit. Using the nanostructuring technique, the ZT values of many materials which were known for decades were improved and values close to 2 and above were reported for some of these materials (Biswas et al. 2012; LaLonde et al. 2011; Liu et al. 2012; Zhao et al. 2014).

Future of Solid-State Heat Engines

Thermoelectricity is an old field and started back in 1821 when Seebeck first observed the deflection of the magnetic needle as a result of heating of his circuit. There are many contributors to the field, and it is beyond the scope of this writing to name all of the scientists who contributed to different aspects, from fundamentals to devices and applications. Yet thermoelectric modules do not have a large market.

Today, power plants can burn fuel and generate electricity with efficiencies as high as 30–50%. But thermoelectric power generators available in the market are still less than 10% efficient. Many of the good thermoelectric materials found have large ZT values at high temperatures but low values at room temperatures. The peak ZT is not what determines the efficiency, but rather it is the average ZT. Therefore, scientists are still in search of good thermoelectric materials with average ZT values above 1 over a wide temperature range. Many classes of materials have been studied, but many other classes remain to be studied. Heavily doped semiconductors were overwhelmingly the subject of many investigations (Mahan 1998; Chasmar and Stratton 1959; Nolas et al. 2001). However, semimetals are also potential candidates, and they have been overlooked. Semimetals have large electrical conductivity with minimum requirements for doping. While they do not have a bandgap, what is needed for a large Seebeck coefficient is not necessarily a large bandgap, but rather a large asymmetry between electrons and holes. Many semimetals have a large effective mass difference between electrons and holes, and that is enough to achieve a relatively large Seebeck coefficient in these materials (Markov et al. 2018). Another large class of materials that have recently emerged are two-dimensional materials such as graphene and transition metal dichalcogenide layers. These 2D materials could have very large thermoelectric power factor values. They have only been recently studied, and already they have shown record high power factor values (Duan et al. 2016).

Cost and scalability are other big challenges. Current thermoelectric modules are mostly made for small-scale applications and are not scalable to be used in large areas such as power plants or for waste heat recovery in factories. They are mostly made out of bismuth telluride components which are considered expensive rare elements (Yazawa and Shakouri 2013).

Finally, other than power generation and refrigeration, thermoelectric materials can have other applications. They are currently widely used as thermal sensors (thermocouples). But they can also be used for applications such as electronic cooling where heat is pumped from the hot chip to the cold ambient. For such applications, traditional high-ZT materials are not useful. Instead materials with a high thermoelectric power factor and a high thermal conductivity are needed (Zebarjadi 2015). One can define an effective active thermal conductivity for thermoelectric materials working under applied current, which is the measure of their capacity to transport heat. The active thermal conductivity is additive to the passive thermal conductivity and is proportional to the thermoelectric power factor. Unfortunately, the power factor values reported so far are very small. One of the highest bulk thermoelectric power factors is reported for YbAl_3 (Mahan 1998) which is only about $\sigma S^2 T = 5 \frac{W}{mK}$. This value is much smaller compared to [passive] thermal conductivity of a simple copper heat sink $\left(385 \frac{W}{mK}\right)$. Therefore, current thermoelectric materials are not suitable for electronic cooling applications. However, note that materials with high power factor and high thermal conductivity have not been the subject of thermoelectric research in the past and much focus has been on high-ZT materials. Therefore, there is a need to investigate this new class of materials.

Another important application is in thermal switching. In principle, a thermoelectric module could be used to switch from weak heat conduction to large heat conduction in its off-on modes. This last application for thermoelectric modules has only been investigated this year (Adams et al. 2019).

Solid-state thermionic devices were proposed in 1997 (Shakouri and Bowers 1997; Mahan et al. 1998), but there has been only limited demonstration of these devices. The field of SSTIs is still considered new and open. There are many materials, especially in the layered material family, that one can investigate for thermionic applications. These are good for small-temperature gradients and at very small scales. They can also provide a very good solution for electronic cooling.

Acknowledgements I would like to thank Pamela Norris for inviting me to write this chapter and for supporting my research during my time at University of Virginia. I would also like to thank Vivian Lin, who is currently an undergraduate student at UVA for providing feedback on the clarity of the manuscript. This work is supported by National Science Foundation, Career award number 1653268.

References

Adams, M. J., et al. (2019). High switching ratio variable-temperature solid-state thermal switch based on thermoelectric effects. *International Journal of Heat and Mass Transfer, 134,* 114–118.

Biswas, K., et al. (2012). High-performance bulk thermoelectrics with all-scale hierarchical architectures. *Nature, 489*(7416), 414–418.

Borgnakke, C., & Sonntag, R. E. (2013). *Fundamentals of thermodynamics.* Hoboken, NJ: Wiley.

Chasmar, R. P., & Stratton, R. (1959). The thermoelectric figure of merit and its relation to thermoelectric generators. *Journal of Electronics and Control, 7*(1), 52–72.

Chen, G. (2011). Theoretical efficiency of solar thermoelectric energy generators. *Journal of Applied Physics, 109*(10), 104908.

Curtin, B. M., et al. (2013). Field-effect modulation of thermoelectric properties in multigated silicon nanowires. *Nano Letters, 13*(11), 5503–5508.

DOE. (2018). DOE-EIA monthly energy review. *U.S. Energy Information Administration.* Retrieved from https://www.eia.gov/totalenergy/data/monthly/

Duan, J., et al. (2016). High thermoelectric power factor in graphene/hBN devices. *Proceedings of the National Academy of Sciences of the United States of America, 113*(50), 14272–14276.

Goldsmid, H. J. (2010). *Introduction to thermoelectricity.* Berlin Heidelberg: Springer Series in Materials Science.

Hatsopoulos, G., & Kaye, J. (1958). Analysis and experimental results of a diode configuration of a novel thermoelectron engine. *Proceedings of the IRE, 46*(9), 1574–1579.

Heremans, J. P., Wiendlocha, B., & Chamoire, A. M. (2012). Resonant levels in bulk thermoelectric semiconductors. *Energy & Environmental Science, 5*(2), 5510.

Hicks, L., & Dresselhaus, M. (1993a). Effect of quantum-well structures on the thermoelectric figure of merit. *Physical Review B, 47*(19), 12727–12731.

Hicks, L., & Dresselhaus, M. (1993b). Thermoelectric figure of merit of a one-dimensional conductor. *Physical Review B, 47*(24), 16631–16634.

Horst, R. B. (1963). Thermomagnetic figure of merit: Bismuth. *Journal of Applied Physics, 34*(11), 3246.

Hung, N. T., et al. (2016). Quantum effects in the thermoelectric power factor of low-dimensional semiconductors. *Physical Review Letters, 117*(3), 36602.

Ioffe, A. F. (1957). *Semiconductor thermoelements, and thermoelectric cooling.* Infosearch, ltd.

Ioffe, A. F. (1958). The revival of thermoelectricity. *Scientific American,* 199.

Kim, K., & Kaviany, M. (2016). Thermal conductivity switch: Optimal semiconductor/metal melting transition. *Physical Review B, 94*(15), 155203.

Koeck, F. A. M., et al. (2009). Thermionic electron emission from low work-function phosphorus doped diamond films. *Diamond and Related Materials, 18*(5–8), 789–791.

LaLonde, A. D., Pei, Y., & Snyder, G. J. (2011). Reevaluation of PbTe1−xIx as high performance n-type thermoelectric material. *Energy & Environmental Science, 4*(6), 2090.

Liu, H., et al. (2012). Copper ion liquid-like thermoelectrics. *Nature Materials, 11*(5), 422–425.

MacKay, J. C. (2010). *Sustainable energy—Without the hot air.* Cambridge, UK: UIT Cambridge Ltd. (D. Hafemeister, Reviewer).

Mahan, G. (1998). Good thermoelectrics. *Solid-State Physics, 51,* 81.

Mahan, G. D., Sofo, J. O., & Bartkowiak, M. (1998). Multilayer thermionic refrigerator and generator. *Journal of Applied Physics, 83*(9), 17.

Marder, M. P. (2010). *Condensed matter physics.* New York: Wiley.

Markov, M., et al. (2018). Semi-metals as potential thermoelectric materials: Case of HgTe. *Scientific Reports, 8*(1).

Massicotte, M., et al. (2016). Photo-thermionic effect in vertical graphene heterostructures. *Nature Communications, 7,* 12174.

Muller, R. (2008). *Physics for future presidents: The science behind the headlines*. New York: W.W. Norton.

Nolas, G. S., Sharp, J., & Goldsmid, H. J. (2001). *Thermoelectrics: Basic principles and new materials developments*. Berling: Springer.

Poudel, B., et al. (2008). High-thermoelectric performance of nanostructured bismuth antimony telluride bulk alloys. *Science (New York, N.Y.), 320*(5876), 634–638.

Rowe, D. M. (2005). *Thermoelectrics handbook: Macro to nano*. Boca Raton: CRC Press.

Shakouri, A., & Bowers, J. E. (1997). Heterostructure integrated thermionic coolers. *Applied Physics Letters, 71*(9), 1234.

Uchida, K., et al. (2008). Observation of the spin Seebeck effect. *Nature, 455*(7214), 778–781.

Wang, X., Zebarjadi, M., & Esfarjani, K. (2016). First principles calculations of solid-state thermionic transport in layered van der Waals heterostructures. *Nanoscale, 8*(31), 14695–14704.

Wang, X., Zebarjadi, M., & Esfarjani, K. (2018). High-performance solid-state thermionic energy conversion based on 2D van der Waals heterostructures: A first-principles study. *Scientific Reports, 8*(May), 1–9.

Wehmeyer, G., et al. (2017). Thermal diodes, regulators, and switches: Physical mechanisms and potential applications. *Applied Physics Reviews, 4*(4), 41304.

Wilson, V. C. (1959). Conversion of heat to electricity by thermionic emission. *Journal of Applied Physics, 30*(4), 475–481.

Yazawa, K., & Shakouri, A. (2013). Cost-performance analysis and optimization of fuel-burning thermoelectric power generators. *Journal of Electronic Materials, 42*(7), 1946–1950.

Zebarjadi, M. (2015). Electronic cooling using thermoelectric devices. *Applied Physics Letters, 106*(20), 203506.

Zebarjadi, M. (2019). *E-Snail*. http://faculty.virginia.edu/E-Snail/

Zebarjadi, M. (2017). Solid-state thermionic power generators: An analytical analysis in the non-linear regime. *Physical Review Applied, 8*(1), 14008.

Zebarjadi, M., et al. (2011). Power factor enhancement by modulation doping in bulk nanocomposites. *Nano Letters, 11*(6), 2225–2230.

Zebarjadi, M., et al. (2012). Perspectives on thermoelectrics: From fundamentals to device applications. *Energy and Environmental Science, 5*(1), 5147–5162.

Zhao, L.-D., et al. (2014). Ultralow thermal conductivity and high thermoelectric figure of merit in SnSe crystals. *Nature, 508*(7496), 373–377.

Ziabari, A., et al. (2016). Nanoscale solid-state cooling: A review. *Reports on Progress in Physics, 79*(9), 95901.

Chapter 7
Manipulating Water and Heat with Nanoengineered Surfaces

Yangying Zhu, Heena K. Mutha, Yajing Zhao, and Evelyn N. Wang

Introduction

Water is a key element in numerous applications impacting our daily lives. The majority of power plants rely on steam cycles to produce electricity and a significant amount of water (40% of total freshwater withdrawal in the USA) to transfer waste heat to the environment. Central building heating and cooling systems typically use water as a medium to transport heat between the chiller or heat pump, and the indoor and outdoor environment. Desalination technologies convert brackish water and seawater to freshwater for agricultural, industrial, and residential use. In addition, many electronic devices such as central processing units (CPUs), power electronics, batteries in electronic vehicles, and equipment in data centers require liquid cooling or liquid-to-vapor phase change processes to dissipate heat.

The use of water in these applications is closely connected to energy. In most cases, water serves as a medium to transport thermal energy because of its high specific heat capacity (4.184 kJ/kg) or latent heat of vaporization (2260 kJ/kg) as water changes phase from liquid to vapor or vice versa (Lienhard and Lienhard 2003). Improving the thermal transport or heat transport processes in power plants has a significant impact on the efficiency of electricity generation. A 1% increase in overall efficiency achieved by enhancing condensation means producing 10^{17} J more electrical energy per year in the USA, which is equivalent to the energy from 10^7 tons of coal (Wiser 2000). For thermal management of electronics, buildings, or industrial applications, improved heat transfer reduces both energy consumption and water usage. On the other hand, the high latent heat also causes processes that rely on phase change for products such as desalination based on thermal evaporation to be relatively energy intensive. With water and energy as limited resources, it is important to design better thermal systems in these applications.

Y. Zhu · H. K. Mutha · Y. Zhao · E. N. Wang (✉)
Massachusetts Technology of Institute, Cambridge, MA, USA
e-mail: enwang@mit.edu

© Springer Nature Switzerland AG 2020
P. M. Norris, L. E. Friedersdorf (eds.), *Women in Nanotechnology*, Women in Engineering and Science, https://doi.org/10.1007/978-3-030-19951-7_7

Historically, efforts to increase thermal energy utilization and transport have mainly focused on system-level design in thermodynamic cycles to recover waste heat, as well as on increasing the heat transfer area or liquid mixing at the component level (e.g., heat exchangers with fins and grooves). Recently, advances in nanotechnology (i.e., the ability to create structures, material patterns, and interfaces with nanometer-scale resolution and precise control) have introduced tremendous opportunities in various applications such as integrated circuits on semiconductors, sensors and actuators, drug delivery, batteries for energy storage, and solar cells, among many others. Likewise, manipulating water processes in desalination and liquid-vapor phase change with nanotechnology has also attracted significant attention. For desalination, researchers have been able to design and engineer membranes (filtration based) or electrodes (electrochemical method based) with better control over their topography and chemical properties. For liquid-vapor phase change heat transfer, engineered surfaces with micro/nanostructures and chemical composition can control the liquid-vapor interface where phase change occurs, as well as liquid and vapor transport near the interfaces. These nanoengineering approaches have been shown to significantly enhance the heat transfer performance in thermal management applications. In this chapter, we review some of the key progress and contributions we have made for applications in thermal management of electronics, condensation heat transfer for power generation, and water desalination, and also provide an outlook for future directions and opportunities.

Electronics Cooling

The increasing power density of modern electronic devices such as central processing units (CPUs) in commercial computers, power amplifiers, laser diodes, and transceivers poses a significant thermal management challenge. The expected lifetime and efficiency of electronic devices typically reduce as the temperature at which they operate increases. Overheating can cause safety issues in energy storage systems such as batteries. In addition to reliability, safety, and efficiency issues, large-scale facilities that house computers and telecommunication/storage systems (data centers) also consume significant amounts of energy for cooling (liquid pumping, fans, air conditioning, etc.). Effective and efficient electronics cooling schemes that can dissipate a high heat flux, maintain a low device temperature, and require minimal power consumption are highly desired for current and future electronic devices.

Cooling schemes can be categorized into air cooling, single-phase liquid cooling, and liquid-to-vapor phase-change cooling with the end goal of dissipating heat generated from the electronics to the ambient environment. Air cooling removes heat with airflow either through natural convection or forced convection with a fan. Because of the low specific heat of air (1 kJ/kg K at room temperature), air cooling is only applicable to low-heat-flux applications. Meanwhile, liquids have much higher specific heats compared to air, so single-phase liquid cooling can dissipate a higher heat flux with the same temperature rise as air cooling, especially when circulating at high flow rates. Furthermore, the liquid-to-vapor phase change process harnesses the latent heat of vaporization which is even higher than the specific heat for many liquids. For example, for water, the latent heat is 2260 kJ/kg whereas the specific heat capacity is only 4.184 kJ/kg K. As a result, thermal management for high-heat-flux applications using phase change has attracted significant attention. A tremendous amount of research has investigated immersion boiling, flow boiling in mini- and microchannels, and heat pipes where evaporation and condensation occur in wick structures. It is to be noted that for liquid cooling and phase-change cooling, however, the heat from the electronic device is dissipated first into the cooling fluid and eventually into the ambient air, so a complete heat transfer analysis needs to consider the subsequent heat transfer from the fluid to air side, which can lead to high device temperatures if not well designed.

As the heat fluxes in modern electronic devices exceed 100 W/cm^2 or even 1000 W/cm^2, the challenges for liquid-to-vapor-phase-change heat transfer mainly include the need to enhance the critical heat flux (CHF) and to achieve a high heat transfer coefficient (HTC) (Cho et al. 2016). CHF is the maximum heat flux that can be dissipated beyond which heat transfer severely degrades due to liquid dry-out. Thus, it represents the limit to the heat flux that can be dissipated during device operation. HTC is the ratio of the dissipated heat flux over the resulting temperature difference (usually between the solid wall and the bulk fluid). A high HTC is desired because it is associated with a low temperature of the electronic device when the fluid temperature is constant. The key to achieving high CHF is to suppress liquid dry-out on the heated solid surface. A high HTC can be obtained by facilitating vapor nucleation, creating a thin-liquid-film region where the thermal resistance is very low, and enhancing convective heat transfer.

A few different types of liquid-to-vapor heat transfer processes include pool boiling (boiling inside a stationary pool of liquid), flow boiling (boiling as liquid flows through a channel), and evaporation (directly from the wall without bubble generation). Compared with pool boiling and evaporation, flow boiling is especially promising because of its compact form factor and the potential to achieve a high CHF (forced liquid flow delays dry-out), high HTC (forced convection combined with boiling), and uniform temperature. However, flow boiling in mini- and microchannels typically suffers from flow instability which causes severe fluctuations in temperature and pressure drop during operation (Kandlikar 2002; Kakac and Bon 2008). Flow instabilities can be triggered by several mechanisms including explosive bubble expansion (Hetsroni et al. 2005), upstream compressibility (Zhang et al. 2010), and density wave oscillation (Yadigaroglu and Bergles 1972). The temperature and pressure drop oscillations are detrimental for the electronic devices, whose performance is often temperature sensitive, and their lifetime may be severely reduced due to the associated mechanical stress and crack. Therefore, tremendous efforts have been made to suppress flow instabilities during flow boiling. Past efforts mainly include adding an inlet restrictor or valve which introduces a stable single-phase flow resistance that prevents reverse flow from the subsequent two-phase channel. This approach has demonstrated to be effective, but at the cost of the additional pressure drop across the valve which increases the pumping power consumption.

The advancement of nanotechnology with the ability to precisely engineer surface structures as well as manipulate wettability has created opportunities to address the challenges of flow boiling at the microscopic level. One reason for flow instabilities is the lack of nucleation sites, which causes the onset of nucleation to occur at a high wall temperature, and the flow alternates between explosive boiling and single-phase flow. The use of cavities that help to trap gas to facilitate bubble nucleation at a lower wall temperature was proposed to suppress flow instability (Fig. 7.1a). Kuo et al. incorporated well-defined artificial nucleation cavities on the sidewalls of five parallel microchannels (Kuo and Peles 2008) and performed subcooled flow boiling experiments (inlet water temperature at room temperature). The cavities suppressed explosive bubble growth instability compared to plain microchannels. Recently, several works integrated silicon nanowires onto microchannel wall surfaces (Li et al. 2012; Yang et al. 2014a, b). The nanowire bundles formed cavities (a few microns in dimension) and similarly nucleate boiling occurred at a lower temperature. Li et al. showed that the nanowires reduced temperature fluctuations from approximately 15 to 10 K at a relatively low heat flux of 80 W/cm² (mass flux is 571 kg/m² s) and the pressure drop fluctuations were reduced (2–1 psi) as well (Li et al. 2012).

Due to the significant difference in density between liquid and vapor, flow boiling is accompanied with a large-volume expansion of the fluid which leads to flow instability. Therefore, another means to mitigate flow instability is to locally vent the generated vapor from the channel (Fig. 7.1b) to reduce volume expansion. For example, one of the microchannel surfaces is replaced with a hydrophobic membrane that is permeable to vapor but not liquid. David et al. incorporated a hydrophobic polytetrafluoroethylene (PTFE) membrane with 220 nm pores into a microchannel heat exchanger (David et al. 2011a, b) and found that the vapor venting design significantly reduced the pressure drop across the channel and improved the HTC. Fazeli et al. removed the liquid exit of the channel and incorporated pin fins inside the channel which support the membrane and also enhance the heat transfer area (Fazeli et al. 2015). Woodcock et al. fabricated "piranha pin fins" into the channel which serve as nucleation sites and in addition designed vapor venting ports directly above each nucleation site to remove bubbles locally. This combined design achieved a significant heat flux of over 700 W/cm² using refrigerant HFE-7000 as the working fluid.

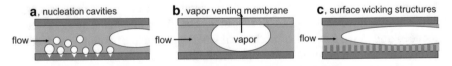

Fig. 7.1 Schematics of flow boiling enhancement via nanoengineered structures. (**a**) Nucleation cavities to promote boiling, (**b**) vapor venting porous membrane to locally remove vapor, and (**c**) surface structures to promote capillarity and wicking

At high heat fluxes, the increased rate of vapor generation results in an annular flow pattern where a vapor core is surrounded by the thin liquid film on the channel wall. Preventing dry-out of the liquid film during mass flow rate oscillations is the key to reducing temperature oscillations and the coupled pressure drop oscillations. Zhu et al. integrated micropillar arrays (Fig. 7.1c) that were optimized for capillary wicking (Zhu et al. 2016b) on the heated surface of a microchannel (Zhu et al. 2016a). At a moderate mass flux (300 kg/m² s), the micropillars promoted capillary flow and significantly suppressed liquid dry-out in the annular flow regime (Zhu et al. 2017). Temperature oscillations were significantly reduced and a 50% enhancement of the CHF was achieved, compared to the smooth surface microchannel, while the pressure drop maintained the same.

In summary, significant efforts have focused on creating micro- and nanostructures to suppress flow instability and enhance CHF in flow boiling for electronics cooling. Several approaches include incorporating nucleation sites with microcavities or nanostructures, reducing fluid volumetric expansion through vapor venting ports or membrane, and suppressing liquid film dry-out by hydrophilic surface structures that promote wicking. These works have demonstrated the promise and feasibility of using nanotechnology to enhance flow boiling and suggest exciting opportunities for thermal management of future high-performance electronic devices.

Condensation for Energy Applications

Condensation of water, the change of phase from vapor to liquid, is commonly observed in nature: clouds, rain, rainbow, fogs, etc. Moreover, condensation is a useful process for energy conversion owing to the large latent heat released during phase change. Condensation of water has been widely utilized in various applications such as steam cycles for power generation, desalination, water harvesting, and nuclear reactors. In these applications, enhancing the efficiency of condensation heat transfer can significantly contribute to energy efficiency, as well as the economic and environmental sustainability of the overall system.

Improving condensation heat transfer has been an ongoing effort for over a century. Conventionally, condensation occurs in the filmwise mode where the condensed liquid forms a film covering the condenser surface. The HTC of filmwise condensation is generally low because the thick liquid film acts as a thermal barrier between the vapor and the condenser surface. A milestone is the discovery of dropwise condensation (Schmidt et al. 1930), where the condensed water forms discrete drops on the condenser surface. The drops grow to a critical size close to the capillary length (2.7 mm for water) and roll off the surface by gravity. In this mode, the surface not covered by droplets is exposed to vapor and therefore heat transfer is more effective. As a result, dropwise condensation has been demonstrated to exhibit significantly higher HTCs compared to filmwise condensation (Rose 2002).

Dropwise condensation is generally achieved by applying a hydrophobic coating to the condenser surface. Nanotechnology has made it possible to explore more designs of surface chemistry and structure, leading to new heat transfer behavior and enhancement. Boreyko and Chen first reported that condensed droplets with diameters much smaller (20–200 µm) than the capillary length (2.7 mm) can jump off a superhydrophobic surface with water contact angles above 150° (Boreyko and Chen 2009). This jumping-droplet behavior is due to conversion of excess surface energy during coalescence to kinetic energy. Miljkovic et al. built on this concept (Fig. 7.2a) to demonstrate a 30% enhancement in the HTC compared with state-of-the-art hydrophobic condensing surfaces using a silanized nanostructured copper oxide surface (Miljkovic et al. 2013).

While the ability to enhance heat transfer with these novel nanoengineered surfaces is promising, a continued challenge with hydrophobic and nanostructured superhydrophobic surfaces is robustness and longevity of the hydrophobic coating. Many engineered surfaces lose hydrophobicity over continuous condensing conditions due to degradation of surface structures and coating (Boinovich and Emelyanenko 2008). As a result, the heat transfer mode changes from dropwise to filmwise and the HTC significantly reduces. Preston et al. recently investigated graphene coatings (Fig. 7.2b) on copper-condensing tubes (Preston et al. 2015). Graphene is an atomic-scale hexagonal lattice composed of carbon atoms, where the advancing contact angle of water on graphene is around 90°. By growing a uniform, single layer of graphene on the copper sample tube through chemical vapor deposition methods, the resulting coated sample demonstrated dropwise condensation with a four times higher condensation HTC compared with filmwise condensation. Although the heat transfer enhancement offered by the graphene coating is slightly lower than the reported 5–7 times enhancement provided by typical hydrophobic coatings, for example, a fluorocarbon monolayer or polymer, graphene coatings showed superior longevity and robustness compared with conventional hydrophobic coatings. This work suggests a promising avenue to promote dropwise condensation for scalable energy applications.

Although dropwise condensation of water vapor could be promoted via various surface designs including nanotextured superhydrophobic surfaces and graphene coatings, the properties of these surfaces are limited by their inability to induce dropwise condensation in systems where low-surface-tension fluids such as hydrocarbons and refrigerants are being used. Fluids with low surface tensions are expected to exhibit low contact angles and spread over conventional hydrophobic surfaces, resulting in filmwise condensation. Nevertheless, a recently developed idea of utilizing a thin film of lubricating liquids trapped in micro/nanoporous media to repel impinging fluids, i.e., slippery liquid-infused porous surfaces (SLIPS), as independently reported by Lafuma et al. and Wong et al. (Lafuma and Quéré 2011; Wong et al. 2011) has paved a novel way to achieve dropwise condensation with low-surface-tension fluids. By taking advantage of the chemical homogeneity and physical smoothness of the liquid-liquid interface of SLIPS, dropwise condensation with low-surface-tension fluids such as toluene, octane, hexane, and pentane has been experimentally demonstrated and a heat transfer enhancement up to five times

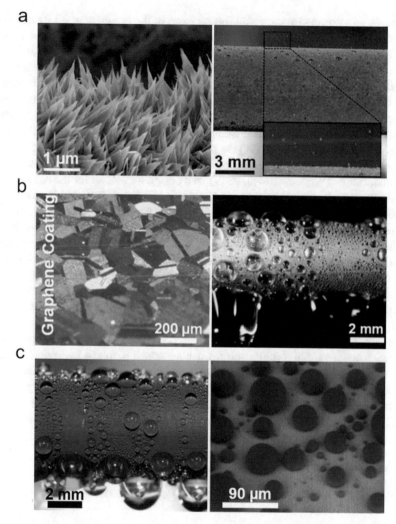

Fig. 7.2 Nanoengineered surfaces to enhance condensation heat transfer. (**a**) Jumping-droplet condensation on silanized nanostructured copper oxide surfaces. (**b**) Dropwise condensation on graphene-coated copper surfaces. (**c**) Dropwise condensation on an oil-infused (tridecafluoro-1,1,2,2-tetrahydrooctyl)-1-trichlorosilane (TFTS)-coated copper oxide surface. Figures reproduced from (**a**) (Miljkovic et al. 2013), American Chemical Society; (**b**) (Preston et al. 2015), American Chemical Society; (**c**) (Xiao et al. 2013), Nature Publishing Group

over filmwise mode has been measured (Rykaczewski et al. 2014; Preston et al. 2018). The exceptional liquid repellency, self-healing ability, and pressure stability exhibited by SLIPS also make it an attractive alternative for those relatively intricate and costly micro- or nanostructured hydrophobic surface designs that can be prone to physical damage and destruction. In a recent study, water condensation on SLIPS (Fig. 7.2c) shows superior performance with a two times higher HTC over state-of-the-art dropwise condensation surfaces in the presence of 30 Pa non-condensable gases (Xiao et al. 2013) and an even higher HTC upon removal of non-condensable gases (Preston et al. 2018). In another study performed by Park et al., a conceptually novel surface design that combines SLIPS with millimeter-sized asymmetric bumpy structures has been successfully applied to dropwise condensation of water vapor and has shown an unprecedented six times higher droplet growth rate compared with other condenser surfaces (Park et al. 2016). In addition to experimental demonstrations on condensation performances of SLIPS, some other studies have focused on understanding the mechanisms behind the performance, such as investigations on wetting behaviors of droplets (Kajiya et al. 2016), droplet growth and distribution (Weisensee et al. 2017), and design guidelines for the combination of condensates, lubricants, and solids (Preston et al. 2017). It should be noted that, although SLIPS has been demonstrated to work for enhancing condensation for various fluids via the dropwise mode, its application remains limited by some intrinsic issues brought by the lubricant layer. For example, the long-lasting functionality of SLIPS suffers from the lubricant drainage due to the "cloaking" behavior (where the lubricant spreads over the condensed droplet to minimize surface energy) or the shear force caused by droplet shedding. Enhancing the robustness of SLIPS against lubricant drainage would be an interesting direction to explore in the future (Sett et al. 2017).

In summary, recent developments of new condensation surfaces aim to sustain dropwise condensation through surface nanostructuring, robust hydrophobic coatings, and lubrication layers. New tools such as laser scanning confocal microscopy and environmental SEM aid the design of such surfaces and understanding of condensation mechanisms. Heat transfer enhancements realized through these nanoengineering approaches can make a large impact on numerous energy and water harvesting applications.

Capacitive Deionization for Water Desalination

At present, over one-third of the world's population lives in water-stressed countries (Elimelech and Phillip 2011), and up-to two-thirds experience water shortages during some part of the year (Mekonnen and Hoekstra 2016). It is predicted that in the next 30 years as global climates warm, glaciers in the Himalayas will recede, severely limiting access to drinking water for an additional 1.5 billion people in India, China, and Southeast Asia (Mayes et al. 2008). Increasing human populations, agricultural and power consumption, and improving living standards will continue to drive up the demand for potable water. While less than 1% of the Earth's water is fresh, over 98% is brackish and sea (Humplik et al. 2011). Therefore, one strategy for increasing water resources is through desalination.

Desalination generates over 37 million cubic meters of potable water per day worldwide, but at present is more expensive than surface water treatment alone. In 2009, the cost of desalination per cubic meter was US $0.53 to US $1.50 for seawater and US \$0.10 to US \$1.00 for brackish water (Greenlee et al. 2009). In order to increase capacity and accessibility, desalination costs need to be reduced. This can be achieved by minimizing the energy consumed and the pretreatment required for production of drinking water. Examining the thermodynamic minimum energy shows that seawater desalination at best consumes 1.1 kWh/m^3, but including process efficiency and overall plant consumption, this may be closer to 3–5 kWh/m^3. Thus, a purification plant delivering 50 L per person daily would require 0.25 kWh per capita. This is only a small fraction of the daily energy consumption in many countries (3.2 kWh per capita in China and 30 kWh per capita in the USA) (Mayes et al. 2008).

There are three main approaches to desalination technology: phase change, separation mechanisms, and charge-based desalination (Humplik et al. 2011). Within these approaches, the most widely used technologies are multistage flash (MSF) and reverse osmosis (RO). In MSF systems, seawater enters low-pressure chambers, where it is flashed to steam and the condensate is collected as potable water. RO is a separation method, where high-pressure water flows across a semipermeable membrane which size-selectively allows water to pass through while rejecting salt. MSF is very inefficient, consuming 25–95 kWh/m^3 or up to 100 times more energy than the theoretical minimum (Humplik et al. 2011), but requires little pretreatment. In contrast, RO typically consumes an average of 5–6 kWh/m^3 for seawater filtration, and as low as 1.6 kWh/m^3 or 40% efficiency (Mayes et al. 2008). However, RO requires water to be pretreated before desalination increasing costs and energy consumption. While MSF and RO are effective desalination systems, they often require large footprints and infrastructure for operation. Charge-based methods for desalination such as capacitive deionization (CDI) could operate more energy efficiently than RO in brackish water streams in a more portable system (Suss et al. 2015). In a typical CDI system, a potential is applied across high-surface-area electrodes. This polarizes the electrode, causing salt ions to adsorb on the surface and desalinate the bulk water flowing through. Following desalination, the potential is removed and ions desorb back into a brine stream, regenerating the cycle.

Nanotechnology enables the development and advancement of these aforementioned technologies. In this section, we focus on work in nanotechnology to enable the development of capacitive deionization for desalination. CDI systems can push desalination capabilities through the design of high-surface-area nanomaterials that have high salt adsorption capacity and fast salt removal rates. Through improvements in desalination realized by nanotechnology, it is possible to create energy-efficient systems to inexpensively meet the rising demand.

Capacitive deionization (CDI) has seen vast technological advancements over the last several years through the design of carbon nanomaterials with surface areas up to 2000 m^2/g. While capacitive deionizaton was first investigated in the 1960s and further developed in the 1980s and 1990s (Blair and Murphy 1960; Oren and Soffer 1983), the system was limited by the surface area of existing carbon materials. Through the recent development of a plethora of activated carbon materials such as mesoporous carbons, aerogels, xerogels, nanotubes, nanowires, and hierarchical carbon materials (Porada et al. 2013; Liu et al. 2015; Suss et al. 2015), salt adsorption capacity has improved from 1–3 mg salt/g carbon up to 26 mg/g (Wimalasiri and Zou 2013). In addition, Porada et al. and others have shown that electrode materials with both macropores and micropores achieve greater storage than those with only macropores (Porada et al. 2012). Finally, tuning surface chemistries to shift the point-of-zero charge to bias the anode and cathode can increase the effective cell voltage and boost desalination and charge efficiency (Cohen et al. 2011).

In addition to material development, CDI has seen engineering breakthroughs (Fig. 7.3a) by modifying the baseline flow-between cell. Membrane-CDI uses ion-exchange membranes over the electrodes to boost charge efficiency and yield higher desalination throughput (Lee et al. 2006). An extension of this has also been to develop flow electrodes, where carbon materials flow in parallel with the saline stream to allow for continuous operation of a CDI system (Jeon et al. 2014; Porada et al. 2014; Hatzell et al. 2014). Finally, Lee et al. have also modified the architecture to incorporate battery-like storage on either one or both electrodes instead of a capacitor to generate a desalination battery to gain salt adsorption capacities up to 31 mg/g (Lee et al. 2014). Figure 7.3b shows salt adsorption capacities achieved by various methods.

However, one challenge in CDI has been understanding how to design flow-between devices coupling system components to porous electrodes while maximizing water quality and throughput. Therefore, Mutha et al. have used electrodes that are straightforward to characterize, vertically aligned carbon nanotubes (VA-CNTs), in order to study and generalize system design with nanomaterials. First, the porosity of varying solid volume fraction VA-CNT forests was designed and characterized, showing that synthesized arrays have inter-CNT spacing of 100 nm, but through mechanical densification can get as close as 10 nm (Mutha et al. 2017). Then a model for maximizing salt rejection for CDI devices with varying volumes,

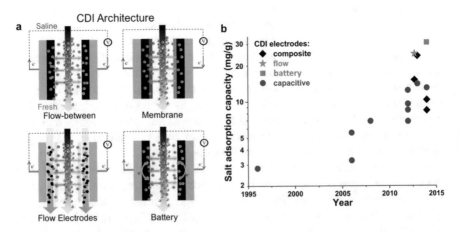

Fig. 7.3 Architectures for capacitive deionization technologies. (**a**) Engineering device architectures such as (**a**) flow-between, membrane, flow electrodes, and battery CDI have enabled (**b**) increasing salt adsorption capacities for realizable desalination. Panel (**b**) adapted with permission from Suss et al. (2015), published by The Royal Science of Chemistry

electrode thicknesses, flow rates, and inlet salt concentrations was developed. The results show that increasing the electrode thickness while reducing the flow channel gap, and operating cells at low flow rates, yielded the highest salt rejection. This approach was demonstrated experimentally using VA-CNT electrodes in a proto-type flow cell (Mutha et al. 2018).

In summary, the design of nanomaterials can advance desalination capabilities in CDI devices. In CDI, selecting porous electrodes that are suited for the device and application can lead to effective desalination. In addition, tuning surface chemistry and engineering device geometry can yield higher, faster throughput.

Summary and Future Directions

In this chapter, we have reviewed some recent advances in thermal management of electronics, condensation for energy, and water desalination enabled by nanotech-nology. For electronics cooling, advances in flow boiling mainly include incorporat-ing nucleation sites with micro-cavities or nanostructures, reducing fluid volumetric expansion through vapor venting ports or membrane, and suppressing liquid film dry-out by hydrophilic surface structures that promote wicking. For condensation, reducing surface wettability to achieve dropwise condensation is the key for HTC enhancement. These are achieved by combining nanostructuring and durable hydro-phobic surface coatings. For water desalination, nanotechnology enables designing membranes or electrodes with more suitable pore size and chemistry for enhanced ion selectivity and energy efficiency.

Despite the significant progress, there is still much room for improvements in current state-of-the-art techniques. First, surfaces that are less susceptible to con-tamination, which can change the surface wettability, are desired as wetting plays a key role in boiling and condensation. This may require design of surfaces that are less likely to adsorb contaminants, self-cleaning surfaces, and methods to desorb contaminants. In addition, robustness of the micro/nanostructures and coatings remains a challenge. Improved bonding between different material layers and reduced stress may help to extend the lifetime of the nanoengineered surfaces. Furthermore, the design of multi-length-scale structures with multi-wettabilities has the potential to achieve higher critical heat fluxes and heat transfer coefficients by having different components of the materials serve different purposes. For example, large microstructures can facilitate capillary liquid flow while small nanostructures enhance thin-film area. For condensation, there is also a need to delay flooding (transition from dropwise to filmwise) on the nanoengineered surfaces when the supersaturation or heat removal rate is high. New designs that will aid faster con-densate removal can potentially address this problem.

In addition to desalination, designing nanomaterials for water harvesting can also yield potable water. Typical systems use dewing or fog capture that requires 100% relative humidity or external energy to drive the system. Recently, we have shown that 1 g of metal organic frameworks (MOFs) can be used at ambient conditions, in sunlight, as low as 20% RH to collect up to 2.8 L water (Kim et al. 2017). This is possible due to the surface chemistry and crystalline structure of MOFs, allowing for water adsorption at typical Earth vapor pressures. These breakthroughs in nanomaterials design allow for novel applications and approaches to collecting potable water. In the future, the smart design of systems using nanoengineering will yield energy-efficient and cost-effective methods to meet the rising demand for water.

References

Blair, J. W., & Murphy, G. W. (1960). Electrochemical demineralization of water with porous electrodes of large surface area. In *Saline water conversion* (pp. 206–223). Memphis, TN: American Chemical Society.

Boinovich, L. B., & Emelyanenko, A. M. (2008). Hydrophobic materials and coatings: Principles of design, properties and applications. *Russian Chemical Reviews, 77*, 583. https://doi.org/10.1070/RC2008v077n07ABEH003775.

Boreyko, J. B., & Chen, C.-H. (2009). Self-propelled dropwise condensate on superhydrophobic surfaces. *Physical Review Letters, 103*, 184501. https://doi.org/10.1103/PhysRevLett.103.184501.

Cho, H. J., Preston, D. J., Zhu, Y., & Wang, E. N. (2016). Nanoengineered materials for liquid–vapour phase-change heat transfer. *Nature Reviews Materials, 2*, 16092. https://doi.org/10.1038/natrevmats.2016.92.

Cohen, I., Avraham, E., Noked, M., et al. (2011). Enhanced charge efficiency in capacitive deionization achieved by surface-treated electrodes and by means of a third electrode. *Journal of Physical Chemistry C, 115*, 19856–19863. https://doi.org/10.1021/jp206956a.

David, M. P., Miler, J., Steinbrenner, J. E., et al. (2011a). Hydraulic and thermal characteristics of a vapor venting two-phase microchannel heat exchanger. *International Journal of Heat and Mass Transfer, 54*, 5504–5516. https://doi.org/10.1016/j.ijheatmasstransfer.2011.07.040.

David, M. P., Steinbrenner, J. E., Miler, J., & Goodson, K. E. (2011b). Adiabatic and diabatic two-phase venting flow in a microchannel. *International Journal of Multiphase Flow, 37*, 1135–1146. https://doi.org/10.1016/j.ijmultiphaseflow.2011.06.013.

Elimelech, M., & Phillip, W. A. (2011). The future of seawater desalination: Energy, technology, and the environment. *Science, 333*, 712–717. https://doi.org/10.1126/science.1200488.

Fazeli, A., Mortazavi, M., & Moghaddam, S. (2015). Hierarchical biphilic micro/nanostructures for a new generation phase-change heat sink. *Applied Thermal Engineering, 78*, 380–386. https://doi.org/10.1016/j.applthermaleng.2014.12.073.

Greenlee, L. F., Lawler, D. F., Freeman, B. D., et al. (2009). Reverse osmosis desalination: Water sources, technology, and today's challenges. *Water Research, 43*, 2317–2348. https://doi.org/10.1016/j.watres.2009.03.010.

Hatzell, K. B., Iwama, E., Ferris, A., et al. (2014). Capacitive deionization concept based on suspension electrodes without ion exchange membranes. *Electrochemistry Communications, 43*, 18–21. https://doi.org/10.1016/j.elecom.2014.03.003.

Hetsroni, G., Mosyak, A., Pogrebnyak, E., & Segal, Z. (2005). Explosive boiling of water in parallel micro-channels. *International Journal of Multiphase Flow, 31*, 371–392. https://doi.org/10.1016/j.ijmultiphaseflow.2005.01.003.

Humplik, T., Lee, J., O'Hern, S. C., et al. (2011). Nanostructured materials for water desalination. *Nanotechnology, 22*, 292001. https://doi.org/10.1088/0957-4484/22/29/292001.

Jeon, S., Yeo, J., Yang, S., et al. (2014). Ion storage and energy recovery of a flow-electrode capacitive deionization process. *Journal of Materials Chemistry A, 2*, 6378–6383. https://doi.org/10.1039/C4TA00377B.

Kajiya, T., Schellenberger, F., Papadopoulos, P., et al. (2016). 3D imaging of water-drop condensation on hydrophobic and hydrophilic lubricant-impregnated surfaces. *Scientific Reports, 6,* 23687. https://doi.org/10.1038/srep23687.

Kakac, S., & Bon, B. (2008). A review of two-phase flow dynamic instabilities in tube boiling systems. *International Journal of Heat and Mass Transfer, 51,* 399–433. https://doi.org/10.1016/j.ijheatmasstransfer.2007.09.026.

Kandlikar, S. G. (2002). Fundamental issues related to flow boiling in minichannels and microchannels. *Experimental Thermal and Fluid Science, 26,* 389–407. https://doi.org/10.1016/S0894-1777(02)00150-4.

Kim, H., Yang, S., Rao, S. R., et al. (2017). Water harvesting from air with metal-organic frameworks powered by natural sunlight. *Science,* eaam8743. https://doi.org/10.1126/science.aam8743.

Kuo, C.-J., & Peles, Y. (2008). Flow boiling instabilities in microchannels and means for mitigation by reentrant cavities. *Journal of Heat Transfer, 130,* 072402–072402. https://doi.org/10.1115/1.2908431.

Lafuma, A., & Quéré, D. (2011). Slippery pre-suffused surfaces. *EPL (Europhysics Letters), 96,* 56001. https://doi.org/10.1209/0295-5075/96/56001.

Lee, J.-B., Park, K.-K., Eum, H.-M., & Lee, C.-W. (2006). Desalination of a thermal power plant wastewater by membrane capacitive deionization. *Desalination, 196,* 125–134. https://doi.org/10.1016/j.desal.2006.01.011.

Lee, J., Kim, S., Kim, C., & Yoon, J. (2014). Hybrid capacitive deionization to enhance the desalination performance of capacitive techniques. *Energy & Environmental Science, 7,* 3683–3689. https://doi.org/10.1039/C4EE02378A.

Li, D., Wu, G. S., Wang, W., et al. (2012). Enhancing flow boiling heat transfer in microchannels for thermal management with monolithically-integrated silicon nanowires. *Nano Letters, 12,* 3385–3390. https://doi.org/10.1021/nl300049f.

Lienhard, J. H., & Lienhard, J. H. (2003). *A heat transfer textbook.* Cambridge, MA: Dover Publications.

Liu, Y., Nie, C., Liu, X., et al. (2015). Review on carbon-based composite materials for capacitive deionization. *RSC Advances, 5,* 15205–15225. https://doi.org/10.1039/C4RA14447C.

Mayes, A. M., Mariñas, B. J., Georgiadis, J. G., et al. (2008). Science and technology for water purification in the coming decades. *Nature, 452,* 301. https://doi.org/10.1038/nature06599.

Mekonnen, M. M., & Hoekstra, A. Y. (2016). Four billion people facing severe water scarcity. *Science Advances,* e1500323. https://doi.org/10.1126/sciadv.1500323.

Miljkovic, N., Enright, R., Nam, Y., et al. (2013). Jumping-droplet-enhanced condensation on scalable superhydrophobic nanostructured surfaces. *Nano Letters, 13,* 179–187. https://doi.org/10.1021/nl303835d.

Mutha, H. K., Lu, Y., Stein, I. Y., et al. (2017). Porosimetry and packing morphology of vertically aligned carbon nanotube arrays via impedance spectroscopy. *Nanotechnology, 28,* 05LT01. https://doi.org/10.1088/1361-6528/aa53aa.

Mutha, H. K., Cho, H. J., Hashempour, M., et al. (2018). Salt rejection in flow-between capacitive deionization devices. *Desalination, 437,* 154–163. https://doi.org/10.1016/j.desal.2018.03.008.

Oren, Y., & Soffer, A. (1983). Water desalting by means of electrochemical parametric pumping. *Journal of Applied Electrochemistry, 13,* 473–487. https://doi.org/10.1007/BF00617522.

Park, K.-C., Kim, P., Grinthal, A., et al. (2016). Condensation on slippery asymmetric bumps. *Nature, 531,* 78–82. https://doi.org/10.1038/nature16956.

Porada, S., Weinstein, L., Dash, R., et al. (2012). Water desalination using capacitive deionization with microporous carbon electrodes. *ACS Applied Materials & Interfaces, 4,* 1194–1199. https://doi.org/10.1021/am201683j.

Porada, S., Zhao, R., van der Wal, A., et al. (2013). Review on the science and technology of water desalination by capacitive deionization. *Progress in Materials Science, 58,* 1388–1442. https://doi.org/10.1016/j.pmatsci.2013.03.005.

Porada, S., Weingarth, D., Hamelers, H. V. M., et al. (2014). Carbon flow electrodes for continuous operation of capacitive deionization and capacitive mixing energy generation. *Journal of Materials Chemistry A, 2,* 9313–9321. https://doi.org/10.1039/C4TA01783H.

Preston, D. J., Mafra, D. L., Miljkovic, N., et al. (2015). Scalable graphene coatings for enhanced condensation heat transfer. *Nano Letters, 15,* 2902–2909. https://doi.org/10.1021/nl504628s.

Preston, D. J., Song, Y., Lu, Z., et al. (2017). Design of lubricant infused surfaces. *ACS Applied Materials & Interfaces, 9,* 42383–42392. https://doi.org/10.1021/acsami.7b14311.

Preston, D. J., Lu, Z., Song, Y., et al. (2018). Heat transfer enhancement during water and hydrocarbon condensation on lubricant infused surfaces. *Scientific Reports, 8,* 540. https://doi.org/10.1038/s41598-017-18955-x.

Rose, J. W. (2002). Dropwise condensation theory and experiment: A review. *Proceedings of the Institution of Mechanical Engineers Part A—Journal of Power and Energy, 216,* 115–128. https://doi.org/10.1243/09576500260049034.

Rykaczewski, K., Paxson, A. T., Staymates, M., et al. (2014). Dropwise condensation of low surface tension fluids on omniphobic surfaces. *Scientific Reports, 4,* 4158. https://doi.org/10.1038/srep04158.

Schmidt, E., Schurig, W., & Sellschopp, W. (1930). Versuche über die Kondensation von Wasserdampf in Film-und Tropfenform. *Technical and Mechanical Thermodynamics, 1,* 53–63. https://doi.org/10.1007/BF02641051.

Sett, S., Yan, X., Barac, G., et al. (2017). Lubricant-infused surfaces for low-surface-tension fluids: Promise versus reality. *ACS Applied Materials & Interfaces, 9,* 36400–36408. https://doi.org/10.1021/acsami.7b10756.

Suss, M. E., Porada, S., Sun, X., et al. (2015). Water desalination via capacitive deionization: What is it and what can we expect from it? *Energy & Environmental Science, 8,* 2296–2319. https://doi.org/10.1039/C5EE00519A.

Weisensee, P. B., Wang, Y., Hongliang, Q., et al. (2017). Condensate droplet size distribution on lubricant-infused surfaces. *International Journal of Heat and Mass Transfer, 109,* 187–199. https://doi.org/10.1016/j.ijheatmasstransfer.2017.01.119.

Wimalasiri, Y., & Zou, L. (2013). Carbon nanotube/graphene composite for enhanced capacitive deionization performance. *Carbon, 59,* 464–471. https://doi.org/10.1016/j.carbon.2013.03.040.

Wiser, W. (2000). *Energy resources—Occurrence, production, conversion, use* (1st ed.). New York: Springer.

Wong, T.-S., Kang, S. H., Tang, S. K. Y., et al. (2011). Bioinspired self-repairing slippery surfaces with pressure-stable omniphobicity. *Nature, 477,* 443–447. https://doi.org/10.1038/nature10447.

Xiao, R., Miljkovic, N., Enright, R., & Wang, E. N. (2013). Immersion condensation on oil-infused heterogeneous surfaces for enhanced heat transfer. *Scientific Reports, 3.* https://doi.org/10.1038/srep01988.

Yadigaroglu, G., & Bergles, A. E. (1972). Fundamental and higher-mode density-wave oscillations in two-phase flow. *Journal of Heat Transfer, 94,* 189–195. https://doi.org/10.1115/1.3449892.

Yang, F., Dai, X., Peles, Y., et al. (2014a). Flow boiling phenomena in a single annular flow regime in microchannels (I): Characterization of flow boiling heat transfer. *International Journal of Heat and Mass Transfer, 68,* 703–715. https://doi.org/10.1016/j.ijheatmasstransfer.2013.09.058.

Yang, F., Dai, X., Peles, Y., et al. (2014b). Flow boiling phenomena in a single annular flow regime in microchannels (II): Reduced pressure drop and enhanced critical heat flux. *International Journal of Heat and Mass Transfer, 68,* 716–724. https://doi.org/10.1016/j.ijheatmasstransfer.2013.09.060.

Zhang, T., Peles, Y., Wen, J. T., et al. (2010). Analysis and active control of pressure-drop flow instabilities in boiling microchannel systems. *International Journal of Heat and Mass Transfer, 53,* 2347–2360. https://doi.org/10.1016/j.ijheatmasstransfer.2010.02.005.

Zhu, Y., Antao, D. S., Chu, K.-H., et al. (2016a). Surface structure enhanced microchannel flow boiling. *Journal of Heat Transfer, 138,* 091501. https://doi.org/10.1115/1.4033497.

Zhu, Y., Antao, D. S., Lu, Z., et al. (2016b). Prediction and characterization of dry-out heat flux in micropillar wick structures. *Langmuir, 32*, 1920–1927. https://doi.org/10.1021/acs. langmuir.5b04502.

Zhu, Y., Antao, D. S., Bian, D. W., et al. (2017). Suppressing high-frequency temperature oscillations in microchannels with surface structures. *Applied Physics Letters, 110*, 033501. https:// doi.org/10.1063/1.4974048.

Chapter 8
Engineering Interfaces at the Nanoscale

Pamela M. Norris and LeighAnn S. Larkin

Introduction

The first electronic computer, built during World War II at the University of Pennsylvania, weighed around 30 tons and occupied nearly 167 m^2 (Penn Engineering 2017). This computer, the Electronic Numerical Integrator and Computer (ENIAC), was said to contain over 18,000 vacuum tubes which were used for processing, and it required cooling by a forced air/air-conditioning system. In 1947, the invention of the transistor and the subsequent invention of the integrated circuit revolutionized electronic devices. Transistors are smaller, consume less energy, and are cheaper to produce than vacuum tubes. Today, a computer processing unit (CPU) can fit on a single integrated circuit, referred to as a microprocessor, with billions of transistors packed into an area of just a few squared centimeters. It is hard to imagine our modern-day lives without the microprocessors ubiquitous throughout devices ranging from watches to high-performance computers.

The drive for miniaturization, i.e., reducing size while increasing performance, in devices is captured well by "Moore's law." This "law" was first observed in 1965 by Gordon Moore, which states that the number of transistors in an integrated circuit doubles roughly once every 2 years (Moore 1965). For the past 50 years, the semiconductor industry has followed this "law" with amazing success due to advances in semiconductor processing, increased understanding of material properties, and pressing economic interests (Moore 2003). In recent years, there have been a number of reports on the "end of Moore's law" (Waldrop 2016; Zhirnov et al. 2003; Lambrechts et al. 2018; Kish 2002) and one predominant issue cited in the predicted downfall of Moore's law is heat dissipation and thermal management.

P. M. Norris (✉) · L. S. Larkin
Department of Mechanical and Aerospace Engineering, University of Virginia,
Charlottesville, VA, USA
e-mail: Pamela@virginia.edu; lsl9hd@virginia.edu

© Springer Nature Switzerland AG 2020
P. M. Norris, L. E. Friedersdorf (eds.), *Women in Nanotechnology*, Women in
Engineering and Science, https://doi.org/10.1007/978-3-030-19951-7_8

While the number of transistors per die continues to increase, the size of the die has remained relatively constant, and as a result the packing density of transistors is rapidly increasing. Proportional to the packing density, the power and thermal densities are also increasing exponentially. High power densities can produce a thermal bottleneck, where the heat generation will increase device temperature and prevent reliable operation of the integrated circuit (Krishnan et al. 2007; Pop and Goodson 2006; Pop et al. 2006b; Pop 2010). To complicate the "power problem" further, there are often places within the integrated circuit, such as near the clock, where heat generation is concentrated, leading to huge temperature gradients across a short distance. Thermal management has often been considered at the chip scale or for the device as a whole. However, the localized hot spots produced within integrated circuits can occur within a few microns. As transistors approach length scales of around tens of nanometers, the hot spots can be confined to even smaller regions. As a result, it becomes necessary to understand how to prevent self-heating at length scales on the order of micro- to nanometers.

Heat Transfer in Solids

To understand how to circumvent self-heating in electronic devices, it is important to first understand heat transport and generation mechanisms within a solid. In a solid, atoms are constantly vibrating at high frequencies (~10 THz) and small amplitudes within the confines of their neighboring atoms, as shown in Fig. 8.1. The vibrations of each atom are coupled to the other atoms in its vicinity through virtue of atomic bonding. The collective vibrations produce quantized lattice waves and one quantum of vibrational energy is the phonon (see Fig. 8.1). Heat is conducted through a solid by phonons, i.e., the quantized lattice vibrations.

At a finite temperature, the atoms are vibrating in many different directions, and as a result there is a spectra of phonons excited within a material. Each phonon can be attributed typical wave properties such as frequency of oscillation, w; a wave vector, k; and a velocity, v. In a 1D row of atoms (see Fig. 8.1), the phonons can only propagate along the line of atoms, similar to a standing wave on a string.

Fig. 8.1 Atoms in equilibrium (dotted circles) are on a square 2D lattice. The atoms are vibrating around their equilibrium positions. A snapshot of the atoms out of equilibrium is shown in the filled circles. These waves of oscillations are phonons

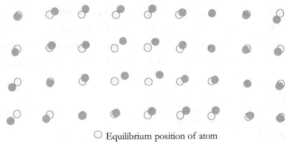

These types of modes are called longitudinal modes, where the displacement of the atoms and propagation of the wave are in the same direction. In three-dimensional solids, the phonons are not restricted to standing waves. Transverse phonon waves, where the displacement and direction of wave motion are perpendicular, are present in materials. These different types of phonon motion are referred to as the polarization.

The phonon frequencies and polarizations which are present in a material depend on the number of atoms in the unit cell, the atoms' arrangement along the lattice points, the atomic masses, and the bond strength between the atoms (Ziman 1967, 2001; Kittel 2001). Calculating the phonon spectra is cumbersome, so researchers use computational tools such as density functional theory or molecular dynamics to input the crystallographic properties, solve the lattice dynamics, and calculate the phonon spectra. The spectra can also be measured experimentally through inelastic neutron scattering.

A phonon spectrum for bulk Ge can be seen in Fig. 8.2. On the left is the dispersion curve. A dispersion curve relates the wave vector to the frequency of the phonon. Each distinct curve is a different polarization. On the right is the phonon density of states (PDOSs). The PDOS describes the number of available states at each available frequency. While the PDOS describes the total number of available

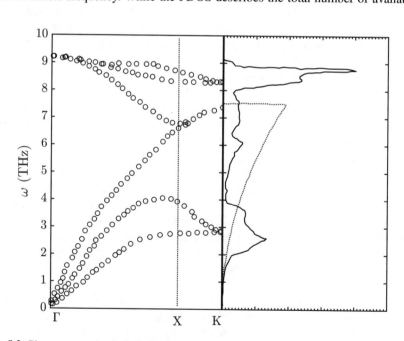

Fig. 8.2 Phonon properties for bulk Ge. On the left is the phonon dispersion; this gives the phonon frequency vs. the wavevector (k). High symmetry points are identified on the dispersion curve. The velocity of the wave is given by the slope of the dispersion. On the right is the phonon density of states (PDOSs). The PDOS presents the total number of states that are available in Ge. The dotted line represents the Debye model for the PDOS. Data is taken from Archilla et al. (2013) (Archilla et al. 2013)

states, the actual occupation of these states is dependent on the temperature of the system. At a finite temperature, the probability that a particular phonon energy is excited (the state is "occupied") is given by the Bose-Einstein function:

$$f_{BE} = \frac{1}{\exp^{\frac{\hbar w}{k_B T}} - 1} \qquad (8.1)$$

where $\hbar$ is the reduced Planck's constant, k_B is the Boltzmann constant, and T is the absolute temperature. At a given temperature, the low-frequency modes, i.e., low-energy phonons, are more likely to be occupied, and as temperature increases more high-energy modes are likely to be excited. This plays an important role in micro-electronics, since the higher the operating temperature, the more phonons partici-pating in transport and scattering.

On the PDOS of bulk Ge, there is a dotted line which is the Debye approximation to the PDOS, which will be referred to later. The Debye model was developed by P. Debye in 1912 to initially describe how phonons contributed to the specific heat of a material (Debye 1912; Davies et al. 1970). It remains a useful first approxima-tion for phonon properties when computational tools and neutron scattering are not available for more precise determination of the PDOS. Debye solids are assumed to have a phonon frequency with a linear relation to the wavevector. In the dispersion curve, the three branches in the dispersion curve that initiate at the origin, which are called the *acoustic* modes, have an approximately linear trend over certain portions of the dispersion. While this model is a first approximation, it can capture certain trends. For example, since the acoustic modes possess the largest velocities, these modes tend to dominate thermal transport. The Debye model also enables the deter-mination of the Debye temperature, Θ_D, the temperature above which all the acous-tic modes are activated within a material.

Electrons can also carry heat, but in semiconductors phonons dominate the heat transport due to the limited mobility of electrons. In semiconducting devices, Electrons carrying current scatter off the phonons, transmitting some of their energy to the phonons. Self-heating arises when the phonons cannot dissipate the heat away fast enough. Hot electrons tend to preferentially scatter off of low-velocity phonon modes (i.e., the optical modes, the modes that in Fig. 8.2 originate at a high frequency at the $k = 0$ point), which further lowers dissipation rates of excess heat (Mahan and Claro 1988; Chen 1996). The result is a material which heats up in localized regions (Rowlette et al. 2005). These regions occur anywhere there is a significant amount of electrical current being conducted or where phonons are likely to be scattered, such as at boundaries and interfaces. Reducing or control-ling the scattering at interfaces would enable quicker dissipation of heat and sup-press the large temperature gradients in localized areas, increasing device reliability and performance. The focus of this chapter is to investigate the fundamental physics of phonon scattering at solid-solid interfaces and methods that can be utilized to reduce the resistance which can lead to self-heating within devices.

Thermal Boundary Conductance vs. Thermal Contact Conductance

Consider heat flowing between two solid materials in contact, as shown in Fig. 8.3. A steady heat current q (in W) across a cross-sectional area A (in m^2) induces a spatial temperature profile ΔT (in K). The temperature gradients in the transport direction in the bulk materials 1 and 2 define their thermal conductivities, κ_i, while the temperature discontinuity at the interface ΔT defines its thermal conductance h, or equivalently its thermal resistance, which is the inverse of the conductance, $R = 1/h$. In macroscopic systems, the conductance is referred to as thermal contact conductance and it is primarily due to imperfect contact between the two surfaces. In such a case, the conductance is limited due to voids or gaps between the two materials, which are macroscopic features. These macroscopic features yield a thermal contact conductance, h_C, typically in the range of 10^4–10^5 W/m^{2}K (Yovanovich 2005; Cooper et al. 1969). As interface quality improves, the conductance increases.

However, even at a "perfect" interface (refer to the right of Fig. 8.3), with the two materials in perfect contact with one another, there remains a small but finite temperature drop across the interface known as the thermal boundary conductance. The thermal boundary conductance, h, arises because the phonons in the material can be partially reflected, even at perfect interfaces. Phonon reflection is primarily caused by differences in the phonon populations in the two materials. The magnitude of the thermal boundary conductance, h, is typically on the order of 10^7–10^9 W/m^{2}K (Hopkins 2013; Monachon et al. 2016).

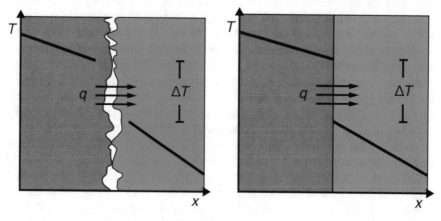

Fig. 8.3 Schematic representation of thermal contact conductance (left) and the thermal boundary conductance (right) at interfaces. Thermal contact conductance, h_C, is limited by the gaps and voids between the two materials. On the right-hand side, the two materials are in perfect contact with one another, and can even be bonded to one another. There is still a temperature discontinuity at this interface caused by differences in the phonon populations in the two materials. Note that these figures are not to scale, since the temperature discontinuity on the left-hand side would be much larger than the discontinuity on the right

In macroscopic applications, the distances between interfaces are large, and the temperature drops at interfaces are negligible compared to those in the bulk materials. However, in many modern devices, the distances between interfaces can be extremely short (tens of nm), and h can become the dominant bottleneck. Shrinking the lengths of the bulk materials would increase the relative importance of the temperature drop ΔT at the interface compared to the bulk materials. For a more concrete example, temperature profiles of AlGaN/GaN heterojunction field-effect transistor (HFET) devices during operation were measured using Raman thermography. The peak temperature of the device is around 240 °C above room temperature. The GaN/SiC interface produces around 30–40% of the temperature rise, while the SiC substrate produces 40–50%. The phonon scattering at the GaN/SiC interfaces produces the same thermal resistance as the entire SiC substrate (Killat et al. 2012; Kuball et al. 2003; Sarua et al. 2007). The interfacial resistance has a tremendous effect in the operation of nanoscale devices and from the perspective of a thermal engineer it becomes imperative to engineer ways to reduce the interfacial resistance.

Controlling Thermal *Contact* Conductance: Carbon Nanotube Case Study

It is possible to enhance h_C through altering the properties of the interface. Thermal interface materials (TIMs) are conductive materials used to fill the voids at the interface between materials to enhance the overall conductance across the interface. In reference to the image on the left of Fig. 8.3, a TIM would be placed in the empty spaces between the hot and cold materials. The most commonly used TIMs on the market are phase-change materials, polymers, thermal greases, and carbon-based materials. The primary requirement for an effective TIM, aside from being thermally and electrically conductive, is to be elastically compliant so that it can support the differing rates of thermal expansion and contraction of the bordering materials (Sarvar et al. 2006). Vertically aligned carbon nanotube arrays (VACNTAs) uniquely possess both the necessary elastic and thermal properties and have been theorized to be an ideal TIM at chip/heat sink interfaces. While several studies have predicted that the thermal conductivity of an individual CNT could reach as high as 3000 W/mK (Pop et al. 2006a; Fujii et al. 2005; Yu et al. 2005), when CNTs are bundled together to form an array the conductivity drops significantly (Bauer et al. 2014). The decrease in conductivity is often attributed to defects caused as a by-product of the array growth process. Much of the work on VACNTAs for TIM applications has been focused on increasing the thermal conductivity of the arrays. These studies often entail an optimization or modification of the growth procedures to minimize irregularities and defects in the tubes (Pham et al. 2017; Bauer et al. 2014).

There has also been an effort to enhance h_C at the interface between the metallic heat sink and the VACNTA. Growth processes, like chemical vapor deposition

(CVD), can leave a canopy of amorphous carbon at the surface of the array which exhibits poor conduction due to its lack of crystallinity and its tendency to not bond well to materials. Recent results by the Norris group demonstrated that in a SiN_x/ VACNTA/Si configuration, the resistance at the VACNTA/interfaces dominates the total resistance to energy flow and can outweigh any increase in the thermal conduction of the array itself (Qui et al. 2017). Qui et al. showed that h_C was largely dependent on the growth conditions for these CVD grown nanotubes, especially the ferrocene (catalyst) sublimation temperature. The contact conductance can be enhanced by a factor of around 3 by varying the sublimation temperature over a range of 50 °C (refer to Fig. 8.4). The sublimation temperature determines how the catalyst particles distribute on the substrate, and as a result how the tubes grow (their length and diameters), as well as the uniformity of the distribution. More uniform distribution of the catalysts occurs at ~137 °C which results in uniform tubes (in height, diameter, and defect density) and minimizes the amount of amorphous carbon by-product, which in turn enhances h_C by a factor of 2. Focusing on minimizing the gaps between the two materials through TIMs is an effective way to mitigate the heating issues from the finite h_C.

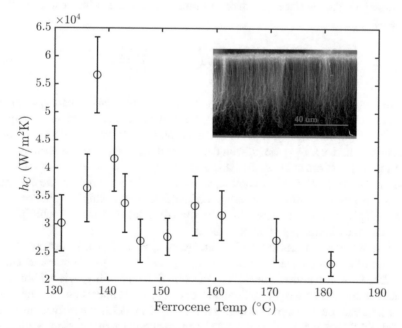

Fig. 8.4 The thermal contact conductance, h_C, of the CNT-solid interface. Varying the sublimation temperature of the ferrocene catalyst impacts the growth process of the CNTs, and as a result can be used to reduce h_C. The inset depicts a micrograph captured on a Quanta 650 scanning electron microscope of a vertically aligned carbon nanotube array grown on a quartz substrate (Qui et al. 2017; Pham et al. 2017)

Phonon-Dominated Thermal Boundary Conductance

Similar to the CNT case described above, it is possible to engineer an interface to enhance the thermal boundary conductance, h. The ability to control h requires the ability to control on the scale of phonon scattering—the nanoscale. In order for thermal engineers to make significant headway in improving h to optimize devices for performance and reliability, the influence of different properties and interface features on h and how they can be controlled must be understood. The following sections present a brief overview of how to predict h at interfaces and some methods that can be used to tune h.

For a given interface, h can be difficult to predict a priori since h is dependent on both characteristics that are fundamental to the materials compromising the interface, such as the PDOS, and extrinsic factors related to characteristics of the interface (bond strength, interfacial roughness, dislocation density, and diffusion at the interface). Thermal boundary conductance is also influenced by external factors such as temperature and pressure. Using a semiclassical formalism, significant insight into some of the factors influencing h can be provided. Most of the semiclassical models are based upon the same expression for an ensemble of phonons impinging on the interface, and their resulting conductance, which can be summarized by

$$h_{\mathrm{BD}} = \frac{1}{4} \frac{j}{3} \sum \int\limits_{w_{D,j}}^{0} \alpha \hbar w v_{1,j} D\left(w, v_{1,j}\right) \frac{df_{\mathrm{BE}}}{dT} \, dw \qquad (8.2)$$

where j represents a phonon branch, w is the angular phonon frequency, $w_{D,j}$ is the maximum frequency, i.e., the Debye frequency for a phonon mode, v_j is the group velocity of phonon mode j, α is the transmission probability of mode j to transmit across the interface, D_j is the phonon density, and f_{BE} is the Bose-Einstein distribution function (Norris et al. 2013). This general expression employs a few assumptions including the Debye assumption, an isotropic dispersion within the materials, a small temperature differential across the interface, and equilibrium phonon populations. This expression shows that h is directly related to the available phonon populations on both sides of the interface.

Most of the parameters in Eq. (8.2), including w, v, and D, can be directly quantified from the phonon spectra which are often available in the literature or can be calculated using ab initio methods. The transmission probability, α, is not obviously related to the phonon spectra and remains one of the least understood parameters in Eq. (8.2). The most commonly used semiclassical model is the diffuse mismatch model (DMM), which assumes that all phonon scattering at the interface is diffusive (Swartz and Pohl 1989). Diffusive scattering implies that after a phonon scatters off the interface, the phonon "does not remember" its original polarization or incident angle. At room temperature and higher, the assumption of a perfectly diffuse interface is fairly accurate. With the diffusive assumption, the transmission probability can be derived from the principle of detailed balance, which states that the interface must be in local equilibrium, i.e., $q_{1-2} = q_{2-1}$:

$$\alpha = \frac{v_{j,2}^{-2}}{v_{j,2}^{-2} + v_{j,1}^{-2}} \tag{8.3}$$

Thus, α is dependent upon the group velocities in the two materials, which is related to the phonon spectra of the two materials. The DMM is the most widely used model for predicting h at a metal/semiconductor interface a priori (Duda et al. 2010). The accuracy of the DMM predictions compared to experimental measurements has received a lot of attention in the literature. There has been a significant amount of work focused on improving DMM predictions by removing some of the assumptions made in the above derivation, i.e., using the full dispersion and/or including the optical branches instead of only using the Debye assumption (Reddy et al. 2005; Beechem et al. 2010). Even still, most often, the DMM tends to overpredict the experimental measurements of h since the DMM does not explicitly account for interfacial imperfections, like defects, roughness, or interdiffusion (Stevens et al. 2005). There have been a myriad of DMM modifications to explicitly account for interfacial imperfections and these modifications have been successful for predicting h at some interfaces (Beechem et al. 2007, Beechem and Hopkins 2009). Occasionally, however, the original DMM has been shown to underpredict the experimental measurements (Lyeo and Cahill 2006; Hopkins et al. 2007). Underpredictions are typically attributed to other scattering mechanisms present at the interface, such as electron-phonon coupling, inelastic (higher order processes) scattering, or interfacial states, and there have been a number of separate DMM modifications to account for these processes (Lyeo and Cahill 2006; Duda et al. 2011; Hopkins and Norris 2009). All of these different models and their accuracy paint the picture of how complicated it can be to predict h at a metal/semiconductor interface.

Phonon Density of States and Thermal Boundary Conductance

Regardless of the assumptions made in the DMM, the original model does capture one significant aspect of h: the dependence upon the similarity of the vibrational properties in the two materials. When the vibrational properties (the velocities) of the phonons in the two materials are similar, the transmission probability approaches a maximum of unity, and conversely, when there is a large mismatch in the vibrational properties, α approaches zero. Visually, we can understand this by looking at the interactions that can occur when a phonon approaches an interface, as shown in Fig. 8.5.

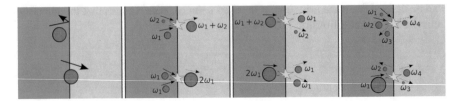

Fig. 8.5 Different types of collisions that a phonon can participate in at an interface. The leftmost image depicts a two-phonon, elastic or harmonic, collision. Each subsequent image depicts different kinds of higher order collisions. These processes are anharmonic or inelastic collisions

It is useful when considering scattering to consider phonons as particles, instead of waves. One possible scattering event occurs when the phonon is reflected or transmitted with the same frequency (energy), as depicted in the leftmost image of Fig. 8.5. This type of collision is called *elastic*, or *harmonic*, since the incident phonon maintains its same energy before and after the collision. An elastic collision is, by far, the most common type of interaction to occur, because it only requires the involvement of one phonon. As we move towards the right in Fig. 8.5, the collisions become more complicated. These *inelastic*, or *anharmonic*, higher order collisions require two or more phonons to participate in the event. The more phonons which are required to participate, the less likely the event is to occur (Hopkins and Norris 2009).

In the elastic limit, the phonon frequency must exist in both materials for the phonon to transmit across the interface. Therefore, the greater the overlap in the phonon frequencies available in the two materials, the more phonons are able to participate in transmission, and the higher the value of h. The trend of vibrational similarity (overlap) not only is predicted in the semiclassical models, but has also been demonstrated repeatedly in the literature, both computationally and experimentally. Computational studies are ideally suited to study this trend since the PDOS in each material can be individually tuned. In 1990, Pettersson and Mahan performed lattice dynamics simulations to study h at the interface of two face-centered cubic (FCC) materials, varying the mass ratios and spring constants to purposely investigate the role of acoustic impedance on h (Pettersson and Mahan 1990). They showed that h was decreased as the relative acoustic impedance increases. One work by Stevens et al. used molecular dynamics to vary $(m_1/m_2)^{1/2}$ of the two materials creating the interface. This mass ratio is effectively equivalent to the ratio of the Debye temperatures of the two materials, Θ_D (Stevens et al. 2007). Stevens showed an exponential increase in h as the ratio of Θ_D approached unity. This gave experimentalists a direct way to test the effect of overlap on h. Lyeo and Cahill, Stoner and Maris, Stevens et al., and Cheaito et al. demonstrated that for a range of metal-semiconductor interfaces, the closer the values of Θ_D of the two materials, the higher the h (Cheaito et al. 2015; Stoner and Maris 1993; Stevens et al. 2005; Lyeo and Cahill 2006).

The Intermediate Layer: The Phonon Bridge

Vibrational overlap is an important characteristic for determining the resulting h, but the vibrational properties are intrinsic to the materials themselves, so the PDOSs cannot be tuned without changing the materials. This can be problematic for device design when our material systems are restricted based on the application. In order to increase the conduction at vibrationally mismatched interfaces, researchers have studied the effect of purposely introducing a distinct intermediate layer with a finite thickness.

Work by the Norris group used molecular dynamics to study the effect of an intermediate layer between two FCC Lennard-Jones materials, which are monoatomic materials. When 2–4 monolayers of material with a mass of the arithmetic mean of the materials on either side are inserted between them, the conductance is enhanced by about 50% at low temperatures. The enhancement, in general, is attributed to "bridging" the disparities in the PDOS between the bulk materials (English et al. 2012). This intermediate layer, when it possesses a mass close to the arithmetic mean (recall that in molecular dynamics simulations, the mass of the atom is closely related to the resulting vibrational spectra), creates a larger overlap in the PDOS of the two materials, resulting in enhanced transmission for harmonic collisions. Similar results were found by a number of other computational studies, although there has been some disagreement over the best way to select the mass of the intermediate layer (Polanco et al. 2017, 2015; Liang and Sun 2012; Stevens et al. 2007).

Smoyer chose to use the $(m_1/m_2)^{1/2} = (\Theta_{D,1}/\Theta_{D,2})^{1/2}$ selection method to experimentally verify the effect of intermediate layers on conduction (Smoyer 2015). A Pt/Ni/Si system was selected, and the thickness of the Ni intermediate layer was varied. In Fig. 8.6, the PDOSs for Pt, Ni, and Si are shown. From this figure, it is evident that Pt and Si have very little overlap in frequencies, so that elastic phonon transmission processes are limited. By adding the Ni layer, there is a "bridge" of intermediate phonon frequencies which can now participate in elastic processes. Smoyer showed that a 5–10 nm thickness of Ni could increase the overall h by a factor of almost 3. Not only do these experimental predictions validate the computational studies, but also these results open up the possibility to engineer a wide range of interfaces.

Of course, there are still a number of unanswered questions when it comes to intermediate layers being used to enhance h. All of the studies mentioned thus far focused primarily on materials with single monoatomic materials, which only possess acoustic modes. Lee et al. demonstrated that the mass, or Debye temperature ratios, for predicting the ideal intermediate layer is not as straightforward when there are optical modes present (Lee and Luo 2017). Due to their complex unit cell, semiconductors often possess a number of optical modes. There has also been some recent work studying the role of an intermediate layer that has a mass or composition grading. This type of grading can occur naturally through the growth processes of semiconducting layers. Zhou et al. showed that an interface that slowly transitioned—layer by layer—from material 1 to 2 can also be used to enhance h (Zhou

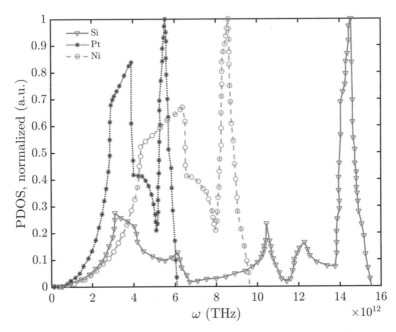

Fig. 8.6 The calculated PDOS for Si (diamonds with solid lines), Pt (closed circles connected with dotted line), and Ni (open circles with dashed lines) derived using Quantum Espresso (Giannozzi et al. 2009). The Ni mediates the vibrational overlap between the Pt and Si. These PDOSs were calculated with Quantum Espresso using a PBESOL potential

et al. 2016). This area of research will continue to be an active and promising path of study for nanoscale thermal scientists to engineer interfaces for enhanced conduction.

Interfacial Quality: Roughness, Diffusion, and Bond Strength

The discussion above has considered the treatment of ideal interfaces, i.e., those without defects, roughness, or diffusion. This is unrealistic for most interfaces. Preparation techniques of substrates can induce roughness on the nanometer length scale, similar length scales to the wavelength of a phonon. Thin-film deposition techniques can create interfaces with poor adhesion and even room temperature can be sufficient to promote interdiffusion between some metals and semiconductors. The quality of the interface can greatly impact the transmission probability, α. For example, the reported values of h at room temperature for the Al/diamond interface range from ~20 to 200 MW/m²K (Monachon and Weber 2013; Monachon et al. 2016). The large range in h on Al/diamond is not atypical of the experimental values of h at metal/semiconductor interfaces, and often searching through the literature one will find several values of h for the same interface. Clearly, there is more than just the overlap in the vibrational spectra which is influencing h. Characteristics at the interface can significantly alter the value of h, and it is thus imperative that these properties are understood. Improving the interfacial quality is one of the most important avenues for engineering interfaces for enhanced phonon conduction. In the next sections, the interfacial characteristics that can influence h will be considered.

Bond Strength/Adhesion and Thermal Boundary Conductance

In the DMM, the metal and semiconductor are assumed to be well bonded, i.e., there is an infinite spring constant connecting the materials. In reality, the bond strength between two materials is finite. Young and Maris first used lattice dynamics in 1989 to investigate the interfacial spring constant, its effect on phonon transmission, and the resulting h (Young and Maris 1989). For weakly bonded interfaces, h increased with the interfacial spring constant, while for strongly bonded interfaces, as defined by a spring constant equal to or greater than the spring constants of the bulk, there was little effect of increasing the spring constant on h. Shen et al. used molecular dynamics to study the interfacial stiffness at weakly and strongly bonded interfaces, using a definition similar to that of Young and Maris, and demonstrated the same trends in bond strength and h (Shen et al. 2011). A number of other computational studies have been performed, all demonstrating that for weakly bonded interfaces, increasing the interfacial stiffness increases the value of h, with the value of h eventually saturating.

Monachon et al. compared density functional theory calculations of the work of adhesion of a metal/semiconductor interface and found h to increase linearly with increasing work of adhesion (Monachon et al. 2014). The adhesion of a metal to semiconductor can be tuned through surface preparation techniques and can subsequently influence h. For example, consider the metal/diamond interface. As-received diamond substrates present lower values of h than diamond substrates than those which are subjected to surface cleaning prior to metal deposition (Monachon et al. 2016). It is theorized that the organic residue left behind on the as-received diamond substrates can strongly disrupt the bond formation between metals and diamond substrate. Sputtering, which tends to promote adhesion, the metal film onto a clean diamond substrate yields higher values of h compared to results with an evaporated metal film (Stevens et al. 2005). Surface termination of diamond has also been demonstrated to increase h. Oxygen termination at the Al/diamond and Cr/diamond interface can increase h by factors of 2 and 11, respectively, compared to the untreated interfaces (Monachon et al. 2014, 2016). These surface preparation techniques to enhance adhesion are not limited to metal/diamond interfaces. Removing the native oxide on Si and SiC through etching is believed to enhance adhesion and has been shown to enhance h at a number of metal/Si or metal/SiC interfaces (Hsieh et al. 2011; Duda and Hopkins 2012; Hopkins et al. 2010). Duda et al. also demonstrated that by inserting a thin, ~7 nm, adhesion layer in a Au/Si system, which has notoriously low adhesion, h can be enhanced by a factor of 4 (Duda et al. 2013). Larkin et al. recently demonstrated a relation between the damping of acoustic waves, which is correlated to the adhesion of the two materials, and h at a variety of metal/In-based semiconductor interfaces (Larkin et al. 2016).

While the above computational results and experimental results demonstrate that, in general, the better bonded (or adhered) a metal is to a substrate, the higher the magnitude of h, there have been very few experimental studies which systematically tune the bond strength at an interface. Hsieh et al., Hohensee et al., and Wilson et al. all used a diamond anvil cell to investigate h as a function of applied pressure at a series of interfaces including Al/SiC, Al/graphene/SiC, Al/MgO, Al/diamond, Pb/diamond, and Pt/diamond (Hsieh et al. 2011; Hohensee et al. 2015; Wilson et al. 2015). Increasing the applied pressure to the system is assumed to create a stronger bond between the layers. At weakly bonded interfaces, like Al/graphene/SiC, h was strongly dependent on pressure and h at strongly bonded interfaces increased weakly and then saturated at high pressures, mirroring the results of Shen et al. and Young and Marris (Shen et al. 2011; Young and Maris 1989).

Functionalization of polymers at weakly bonded interfaces has also been used to systematically investigate the role of bonding on h. Losego et al. measured h at a series of Au/self-assembled monolayers (SAM)/quartz interfaces where the end-group termination of the SAM was varied (Losego et al. 2012). By increasing the number of covalent bonds at the Au/SAM interface, the effective conductance of the system was increased by a factor of 2. A similar approach was used by O'Brien et al. at a Cu/molecule/SiO$_2$ interface and an increase in the baseline h of Cu/SiO$_2$ by a factor of 4 was observed (O'Brien et al. 2013). Hopkins et al. used oxygen termination at Al/single-layer graphene (SLG)/SiO$_2$ to double h (Foley et al. 2015). The use of functionalized inorganic materials has been demonstrated to be an effective way of controlling h and could be useful in the integration of organo-electronics, an emerging field of microtechnology.

Interdiffusion

Interdiffusion between the materials is defined by the compositional gradient across the interface which results in a non-abrupt interface. At these non-abrupt interfaces, there is a finite "mixing layer," quantified as the length over which the species in the two materials mix. Throughout the literature, these interfaces have also been referred to as "diffuse interfaces," "staggered interfaces," or sometimes simply "disordered interfaces." Intermixing disrupts the lattice periodicity, which in turn alters the vibrational properties and/or the propagation of the phonons. Interdiffusion can occur for some materials at room temperature and diffusion is often significantly enhanced with exposure at high temperatures, an environment which is probable during the operation of microelectronic devices.

Work by Hopkins et al. demonstrated that the mixing layer at a Cr/Si interface decreased h. A mixing layer of around ~15 nm reduced h by nearly 40% (Hopkins et al. 2008). The reduction in h appeared to be dependent on both the length of the mixing layer and the rate at which the material transitioned from Cr to Si. To date, the work by Hopkins et al. is the most extensive and comprehensive experimental work on mixing layers at phonon-dominated interfaces. Other researchers have alluded to mixing layers influencing the measured values of h at interfaces, but deconvoluting mixing layer from roughness and varying bond strength can be challenging. Experimental studies accompanied with significant characterization could yield fruitful in understanding how these effects interplay with one another.

There have been a significant number of molecular dynamic investigations on the effect of mixing layers on h. In some cases, mixing layers are shown to increase h, and in other cases a decrease is shown. Molecular dynamic simulations by Twu and Ho, Li and Yang, and Choi et al. all demonstrated a reduction in h with interdiffusion in Lennard-Jones potential/Morse potential materials, Si/Ge, and Al/Si interfaces, respectively (Twu and Ho 2003; Li and Yang 2012; Choi and Roberts 2015). All of these aforementioned studies created the mixing layer through a random selection of intermixing of the species. On the other hand, Liang and Tsai found that with systematic mixing they could create a bi-layered material which increased h. The increase in h is optimized for 4–8 monolayers of mixing, above which the finite thermal conductivity of the mixed layer began to reduce the conduction (Liang and Tsai 2012). Stevens et al. also demonstrated almost a twofold increase in h in an alloyed mixing layer (Stevens et al. 2007). Both studies investigated vibrationally mismatched interfaces, which benefit from an alloyed mixing layer which mediates the vibrational properties. The collection of these results suggests that if the mixing is tailored very specifically, mixing could be used to increase h, but in the case of random diffusion, which is more probable in physical samples, interdiffusion or mixing is likely to decrease h. Using interdiffusion or mixed layers remains a potential method for controlling h and warrants further investigation.

Temperature

In the semiclassical models (Eq. 8.1), the only factor dependent upon temperature is the Bose-Einstein distribution. As a result, the DMM predicts a temperature-dependent trend similar to the lattice specific heat. At low temperatures, i.e., temperatures significantly below the lower Θ_D of the two materials, h presents a T^3 trend. As temperature increases and more phonons are able to participate, h approaches a constant value. This saturation occurs at temperatures close to the lower Θ_D. This temperature-dependent trend for h has been demonstrated for a significant number of metal/semiconductor interfaces. This particular temperature-dependent trend implies that elastic scattering is dominating the transport (Fig. 8.5). However, experimental studies of $h(T)$ at strongly vibrationally mismatched interfaces by Lyeo and Cahill (Pb/diamond and Bi/diamond) and Hopkins et al. (Al/ sapphire) have demonstrated a distinctly different temperature trend (Lyeo and Cahill 2006; Hopkins et al. 2007). These studies showed that at temperatures greater than the lower Θ_D of the system (for these cases, Pb, Bi, or Al), h continued to increase monotonically with temperature. Both studies attributed this increase to inelastic processes participating in interfacial transport which enable phonons in the lower Θ_D material to couple and then transmit into higher frequency phonons in the higher Θ_D material, increasing h by effectively increasing the number of channels. As the temperature continues to grow, more channels become available in the material with the higher Θ_D, and so h continues to grow. The largest contributing processes are the elastic (two-phonon) processes, followed by three-phonon processes (see Fig. 8.5), and each consecutive process has a lower probability of occurring.

Molecular dynamic simulations have also captured the monotonic increase in h with respect to T (Stevens et al. 2007). Work by Saaskilahti et al. used molecular dynamics to spectrally decompose the frequencies which are contributing to h at mismatched interfaces and showed that frequency-doubling or frequency-halving (three-phonon) processes are those most likely to contribute to the inelastic processes (these three phonon processes can be seen in Fig. 8.5) which echoes the argument set forth earlier that the more phonons participating, the less likely that interaction to occur (Saaskilahti et al. 2014). The role of anharmonic scattering and h at high temperatures will remain a crucial area of investigation for operating devices, which can heat up significantly.

Conclusions

As outlined here, thermal boundary conductance, h, is dependent upon several properties, some intrinsic to the materials themselves and others characteristic of the interface itself: the phonon spectra of the two materials, the interfacial roughness, mixing layers, pressure, bond strength, and temperature. Once the influence of each of these properties on the resulting h is well understood, it may, in fact, be possible to use these factors to actually tune h in order to optimize thermal performance. Many of the studies highlighted here attempt to isolate one single characteristic at a time, in order to explore these dependencies. In reality, all of these attributes, phonon density of states, mixing layers, roughness, and many more, are present and influence the phonon transport. To understand the interactions of these properties with one another, a more holistic approach is required. And furthermore, it will also be necessary to examine the impact of any tuning approaches on the resulting electrical conduction or operation of these materials within a device. The potential for engineering the interface for enhanced phonon transmission in order to optimize thermal transport remains a topic of great potential for the nanoscale thermal community.

References

Archilla, J. F. R., Coehlo, S. M. M., Auret, F. D., Nyamhere, C., Dubinko, V. I., & Hizhnyakov, V. (2013). Experimental observation of intrinsic localized modes in Germanium. *Springer Series in Materials Science, 22*, 343–362.

Bauer, M., Pham, Q., Saltonstall, C., & Norris, P. (2014). Thermal conductivity of vertically aligned carbon nanotube arrays: Growth conditions and tube inhomogeneity. *Applied Physics Letters, 105*, 151909.

Beechem, T., & Hopkins, P. E. (2009). *Journal of Applied Physics*, (12), 124301.

Beechem, T., Graham, S., Hopkins, P., & Norris, P. (2007). Role of interface disorder on thermal boundary conductance using a virtual crystal approach. *Applied Physics Letters, 90*, 54104.

Beechem, T., Duda, J., Hopkins, P., & Norris, P. (2010). Contribution of optical phonons to thermal boundary conductances. *Applied Physics Letters, 97*, 2008–2011.

Cheaito, R., Gaskins, J., Caplan, M., Donovan, B., Foley, B., Giri, A., Duda, J., Szwejkowski, C., Constantin, C., Brown-Shaklee, H., Ihlefeld, J., & Hopkins, P. (2015). Thermal boundary conductance accumulation and interfacial phonon transmission: Measurements and theory. *Physical Review B, 91*, 035432.

Chen, G. (1996). Nonlocal and nonequilibrium heat conduction in the vicinity of nanoparticles. *ASME Journal of Heat Transfer, 118*, 539–545.

Choi, C., & Roberts, N. (2015). Contributions of mass and bond energy difference and interface defects on thermal boundary conductance. *AIP Advances, 5*, 097160.

Cooper, M., Mikic, B., & Yovanovich, M. (1969). Thermal contact conductance. *International Journal of Heat and Mass Transfer, 12*, 279–300.

Davies, M., Heisenberg, W. K., & Rutherford, E. (1970). Peter Joseph Wilhem Debye, 1884–1966. *Biographical Memories of Fellows of the Royal Society, 16*.

Debye, P. (1912). Zur Theorie der spezifischen Warmen. *Annalen der Physik, 344*, 789–839.

Duda, J. C., & Hopkins, P. E. (2012). Systematically controlling Kaptiza conductance via chemical etching. *Applied Physics Letters, 100*, 111602.

Duda, J., Hopkins, P., Smoyer, J., Bauer, M., English, T., Saltonstall, C., & Norris, P. (2010). On the assumption of detailed balance in prediction of diffusive transmission probability during interfacial transport. *Nanoscale and Microscale Thermophysical Engineering, 14*, 21–33.

Duda, J., Norris, P., & Hopkins, P. (2011). On the linear temperature dependence of phonon thermal boundary conductance in the classical limit. *Journal of Heat Transfer, 133*, 074501.

Duda, J., Yang, C.-Y., Foley, B., Cheaito, R., Medlin, D., Jones, R., & Hopkins, P. (2013). Influence of interfacial properties on thermal transport at gold:silicon contacts. *Applied Physics Letters, 102*, 081902.

English, T., Duda, J., Smoyer, J., Jordan, D., Norris, P., & Zhigilei, L. (2012). Enhancing and tuning phonon transport at vibrationally mismatched solid-solid interfaces. *Physical Review B, 85*, 035438.

Foley, B., Hernandez, S., Duda, J., Robinson, J., Walton, S., & Hopkins, P. (2015). Modifying surface energy of graphene via plasma-based chemical functionalization to tune thermal and electrical transport at metal interfaces. *Nano Letters, 15*, 4876–4882.

Fujii, M., Zhang, X., Xie, H., Ago, H., Takahashi, K., Ikuta, T., Abe, H., & Shimizu, T. (2005). Measuring the thermal conductivity of a single carbon nanotube. *Physical Review Letters, 95*, 065502.

Giannozzi, P., Baroni, S., Bonini, N., Calandra, M., Car, R., Cavazzoni, C., Ceresoli, D., Chiarottie, G. L., Cococcioni, M., Dabo, I., Dal Corso, A., Fabris, S., Fratesi, G., de Gironcoli, S., Gebauer, R., Gerstmann, U., Gougoussis, C., Kokalj, A., Lazzeri, M., Martin-Samos, L., Marzari, N., Mauri, F., Mazzarello, R., Paolini, S., Pasquarello, A., Paulatto, L., Sbraccia, C., Scandolo, S., Sclauzero, G., Seitsonen, A. P., Somgunov, A., Umari, P., & Wentzcovitch, M. (2009). Quantum Espresso: A modular and open-source software for quantum simulations of materials. *Journal of Physics: Condensed Matter, 21*, 395502.

Hohensee, G., Wilson, R., & Cahill, D. (2015). Thermal conductance of metal-diamond interfaces at high pressure. *Nature Communications, 6*, 1–9.

Hopkins, P. (2013). Thermal transport across solid interfaces with nanoscale imperfections: Effects of roughness, disorder, dislocations, and bonding on thermal boundary conductance. *ISRN Mechanical Engineering, 2013*, 1–19.

Hopkins, P., & Norris, P. M. (2009). Relative contributions of inelastic and elastic diffuse phonon scattering to thermal boundary conductance across solid interfaces. *Journal of Heat Transfer, 131*(2), 22402.

Hopkins, P., Salaway, R., Stevens, R., & Norris, P. (2007). Temperature dependent thermal boundary conductance at Al/Al_2O_3 and Pt/Al_2O_3 interfaces. *International Journal of Thermophysics, 28*, 947–957.

Hopkins, P., Norris, P., Stevens, R., Beechem, T., & Graham, S. (2008). Influence of interfacial mixing on thermal boundary conductance across a Chromium/Silicon interface. *Journal of Heat Transfer, 130*, 062402.

Hopkins, P., Phinney, L. M., Serrano, J. R., & Beechem, T. E. (2010). Effects of surface roughness and oxide layer on the thermal boundary conductance at aluminum/silicon interfaces. *Physical Review B, 82*, 085307.

Hsieh, W.-P., Lyons, A., Pop, E., Keblinski, P., & Cahill, D. (2011). Pressure tuning of the thermal conductance of weak interfaces. *Physical Review B, 84*, 1–5.

Killat, N., Montes, M., Pomeroy, J., Paskova, T., Evans, K., Leach, J., Li, X., Ozgur, U., Morkoc, H., Chabak, K., Crespo, A., Gillespie, J., Fitch, R., Kossler, M., Walker, D., Trejo, M., Via, G., Blevins, J., & Kuball, M. (2012). *IEEE Electron Device Letters, 33*, 366–368.

Kish, L. (2002). End of Moore's law: Thermal (noise) death of integration in micro and nano electronics. *Physics Letters A, 305*, 144–149.

Kittel, C. (2001). *Introduction to solid state physics*. New York: Wiley.

Krishnan, S., Garimella, S., & Mahajan, R. (2007). Towards a thermal Moore's law. *IEEE Transactions on Advanced Packaging, 30*, 462–474.

Kuball, M., Rajasingam, S., Sarua, A., Uren, M., Martin, T., Hughes, B., Hilton, K., & Balmer, R. (2003). *Applied Physics Letters, 82*, 124–126.

Lambrechts, W., Sinha, S., Abdallah, J., & Prinsloo, J. (2018). *Extending Moore's Law through advanced semiconductor design and processing techniques*. Boca Raton: CRC Press.

Larkin, L. S., Redding, M. R., Le, N. Q., & Norris, P. M. (2016). Temperature dependent thermal boundary conductance at metal/indium-based III-V semiconductor interfaces. *Journal of Heat Transfer, 139*, 3.

Lee, E., & Luo, T. (2017). The role of optical phonons in intermediate layer-mediated thermal transport across solid interfaces. *Physical Chemistry Chemical Physics, 19*, 18407–18415.

Li, X., & Yang, R. (2012). Effect of lattice mismatch on phonon transmission and interface thermal conductance across dissimilar material interfaces. *Physical Review B, 24*, 155302.

Liang, X.-G., & Sun, L. (2012). Interface structure influence on thermal resistance across double-layered nanofilms. *Microscale Thermophysical Engineering, 9*, 295–304.

Liang, Z., & Tsai, H.-L. (2012). Reduction of solid-solid thermal boundary resistance by inserting an interlayer. *International Journal of Heat and Mass Transfer, 55*, 2999–3007.

Losego, M. D., Grady, M. E., Sottos, N. R., Cahill, D. G., & Braun, P. V. (2012). Effects of chemical bonding on heat transport across interfaces. *Nature Materials, 11*(6), 502–506.

Lyeo, H.-K., & Cahill, D. (2006). Thermal conductance of interfaces between highly dissimilar materials. *Physical Review B, 73*, 144391.

Mahan, G., & Claro, F. (1988). Nonlocal theory of thermal conductivity. *Physical Review B, 38*, 1963–1969.

Monachon, C., & Weber, L. (2013). Influence of diamond surface termination on thermal boundary conductance between Al and diamond. *Journal of Applied Physics, 113*, 183504.

Monachon, C., Schusteritsch, G., Kaxiras, E., & Weber, L. (2014). Qualitative link between work of adhesion and thermal conductance of metal/diamond. *Journal of Applied Physics, 115*, 123509.

Monachon, C., Weber, L., & Dames, C. (2016). Thermal boundary conductance: A materials science perspective. *Annual Review of Material Research, 8*, 1–31.

Moore, G. (1965). Cramming more components into integrated circuits. *Electronics, 38*, 114–117.

Moore, G. (2003). No exponential is forever, but forever can be delayed. *IEEE International Solid-State Circuits Conference Proceedings, 1*, 20–23.

Norris, P., Le, N., & Baker, C. (2013). Tuning phonon transport: From interfaces to nanostructures. *Journal of Heat Transfer, 135*, 061604.

O'Brien, P.J., Shenogin, S., Liu, J., Chow, P.K., Laurencin, D., Mutin, P.H., Yamaguchi, M., Keblinski, P., & Ramanath, G. (2013). Bonding-induced thermal conductance enhancement at inorganic heterointerfaces using nanomolecular monolayers. *Nature Materials, 12*(2), 118–122.

Penn Engineering. (2017). *ENIAC at Penn Engineering.* Retrieved from www.seas.upenn.edu/about/history-heritage/eniac/.

Pettersson, S., & Mahan, G. (1990). Theory of thermal boundary resistance between dissimilar lattices. *Physical Review B, 42*, 7386–7390.

Pham, Q., Larkin, L., Lisboa, C., Saltonstall, C., Qui, L., Schuler, J., Rupert, T., & Norris, P. (2017). Effect of growth temperature on the synthesis of carbon nanotube arrays and amorphous carbon for thermal applications. *Physica Status Solidi A, 214*, 1600852.

Polanco, C., Rastgarkafshgarkolaei, R., Zhang, J., Le, N., Norris, P., Hopkins, P., & Ghosh, A. (2015). Role of crystal structure and junction morphology on interface thermal conductance. *Physical Review B, 92*, 144302.

Polanco, C. A., Rastgarkafshgarkolaei, R., Zhang, J., Le, N. Q., Norris, P. M., & Ghosh, A. W. (2017). Design rules for interfacial thermal conductance: Building better bridges. *Physical Review B, 95*, 195303.

Pop, E. (2010). Energy dissipation and transport in nanoscale devices. *Nano Research, 3*, 147–169.

Pop, E., & Goodson, K. (2006). Thermal phenomena in nanoscale transistors. *Journal of Electronic Packaging, 128*, 102–108.

Pop, E., Mann, D., Wang, Q., Goodson, K., & Dai, H. (2006a). Thermal conductance of an individual single-wall carbon nanotube above room temperature. *Nano Letters, 6*, 96–100.

Pop, E., Sinha, S., & Goodson, K. (2006b). Heat generation and transport in nanometer-scale transistors. *Proceedings of IEEE, 94*, 1587–1601.

Qui, L., Scheider, K., Radwan, S. A., Larkin, L., Saltonstall, C., Feng, Y., Zhang, X., & Norris, P. (2017). Thermal transport barrier in carbon nanotube array nano-thermal interface materials. *Carbon, 120*, 128–136.

Reddy, P., Castelino, K., & Majumdar, A. (2005). Diffuse mismatch model of thermal boundary conductance using exact phonon dispersion. *Applied Physics Letters, 87*, 211908.

Rowlette, J., Pop, E., Sinha, S., Panzer, M., & Goodson, K. (2005). Thermal simulation techniques for nanoscale transistors. In *IEEE ACM International Conference on Computer-Aided Design*, pp. 225–228.

Saaskilahti, K., Oksanen, J., Tulkki, J., & Volz, S. (2014). Role of anharmonic phonon scattering in the spectrally decomposed thermal conductance at planar interfaces. *Physical Review B, 90*, 134312.

Sarua, A., Ji, H., Hilton, K. P., Wallis, D. J., Uren, M. J., Martin, T., & Kuball, M. (2007). Thermal boundary resistance between GaN and substrate in AlGaN/GaN electronic devices. *IEEE Transactions on Electron Devices, 54*(12), 3152–3158.

Sarvar, F., Whalley, D., & Conway, P. (2006). Thermal interface materials—A review of the state of the art. In *Proceedings of Electronics System Integration Technology Conference*, pp. 1292–1302

Shen, M., Evans, W., Cahill, D., & Keblinski, P. (2011). Bonding and pressure-tuneable interfacial thermal conductance. *Physical Review B, 84*, 1–6.

Smoyer, J. L. (2015). *Local modification to phononic properties at solid-solid interfaces: Effects on thermal transport.* PhD Thesis, University of Virginia.

Stevens, R. J., Smith, A. N., & Norris, P. M. (2005). Measurement of thermal boundary conductance of a series of metal-dielectric interfaces by the transient thermoreflectance technique. *Journal of Heat Transfer, 127*, 315–322.

Stevens, R., Zhigilei, L., & Norris, P. (2007). Effects of temperature and disorder on thermal boundary conductance at solid-solid interfaces: Nonequilibrium molecular dynamic simulations. *International Journal of Heat and Mass Transfer, 50*, 3977–3989.

Stoner, R., & Maris, H. (1993). Kapitza conductance and heat-flow between solids at temperatures from 50 to 300 K. *Physical Review B, 48*, 16373–16378.

Swartz, E., & Pohl, R. O. (1989). Thermal boundary resistance. *Review of Modern Physics, 61*(3), 605.

Twu, C.-J., & Ho, J.-R. (2003). Molecular-dynamics study of energy flow and the Kapitza conductance across an interface with imperfection formed by two dielectric thin films. *Physical Review B, 67*, 1–8.

Waldrop, M. (2016). The chips are down for Moore's law. *Natures News, 530*, 144–147.

Wilson, R., Apgar, B., Hsieh, W.-P., Martin, L., & Cahill, D. (2015). Thermal conductance of strongly bonded metal-oxide interfaces. *Physical Review B, 11*, 1–7.

Young, D., & Maris, H. (1989). Lattice-dynamics calculation of the Kapitza resistance between fcc lattices. *Physical Review B, 40*, 3685–3693.

Yovanovich, M. (2005). Four decades of research on thermal contact, gap, and joint resistance in microelectronics. *IEEE Transactions on Components and Packaging Technologies, 28*, 182–206.

Yu, C., Shi, L., Yao, Z., Li, D., & Majumdar, A. (2005). Thermal conductance and thermopower of an individual single-wall carbon nanotube. *Nano Letters, 9*, 1842–1846.

Zhirnov, V., Cavin, R., Hutchby, J., & Bourianoff, G. (2003). Limits to binary logic switch scaling—A gedanken model. *Proceedings of the IEEE, 91*, 1934–1939.

Zhou, Y., Zhang, X., & Hu, M. (2016). An excellent candidate for largely reducing interfacial thermal resistance: A nano-confined mass graded interface. *Nanoscale, 8*(4), 1994–2002.

Ziman, J. (1967). The thermal properties of materials. *Scientific American, 217*, 181–187.

Ziman, J. (2001). *Electrons and phonons: The theory of transport phenomena in solids.* Oxford.

Chapter 9
National Nanotechnology Initiative: A Model for Advancing Revolutionary Technologies

Celia Merzbacher

Fifteen years after the enactment of the legislation establishing the National Nanotechnology Initiative (NNI) (U.S. Congress 2003), it is worth reflecting on the framework and policies that allowed the program to take root and grow. The NNI is a coordinated suite of activities across the federal government including research, regulation, standards, workforce education, and public outreach. Collectively these activities created a foundation that promoted multidisciplinary research and development (R&D) and enabled the results to be transitioned in a responsible manner to practical application and public benefit. A measure of the NNI's success is the increasing number of products—from medical implants to advanced electronics—that incorporate nanotechnology.

Nanoscale science, engineering, and technology, or nanotechnology, is defined by the NNI as the "understanding and control of matter at the nanoscale, at dimensions between approximately 1 and 100 nm, where unique phenomena enable novel applications" (National Nanotechnology Initiative 2019a). Advances in instrumentation were essential to progress in nanotechnology research and development. In 1981, Gerd Binnig and Heinrich Rohrer at IBM invented the scanning tunneling microscope, which made it possible to image individual atoms and for which they were awarded the Nobel Prize in 1986 (Royal Swedish Academy of Sciences 1986). Binnig, Calvin Quate, and Christoph Gerber went on to invent the atomic force microscope, which allows one to view, measure, and manipulate materials with sub-nanometer precision (Binnig et al. 1986). Perhaps the best known early example of nanoscale control was the demonstration by Don Eigler in 1989, when he spelled IBM by positioning 35 xenon atoms on a nickel surface (IBM 2009).

Nanotechnology emerged from the confluence of advances in the ability to measure and manipulate matter on the scale of atoms and molecules and the recognition of how such capabilities could be used to create new materials, structures, and

C. Merzbacher (✉)
SRI International, Menlo Park, CA, USA
e-mail: Celia.merzbacher@sri.com

© Springer Nature Switzerland AG 2020
P. M. Norris, L. E. Friedersdorf (eds.), *Women in Nanotechnology*, Women in Engineering and Science, https://doi.org/10.1007/978-3-030-19951-7_9

121

processes in diverse fields and sectors. Nanotechnology is distinguished from other areas of materials science and technology by the novel behavior that occurs at nanometer length scales. For example, bulk gold melts at 1064 °C; however, the melting temperature decreases exponentially with particle size below approximately 10 nm diameter (Buffat and Borel 1976; Gao and Gu 2016). In addition to size-related properties that occur at the nanoscale, molecular length scales correlate to many biological features and processes. For example, the width of a DNA molecule is approximately 2 nm. These characteristics open whole new areas of research and development, such as engineered nanoparticles that target and attack disease at the cellular level in the body. At the same time, the novel and sometimes unpredicted properties of nanomaterials extend to their interaction with the body and the environment, posing potential risks.

Advances in materials science and engineering were necessary for nanotechnology to be "born" but other actions and factors were involved in it becoming a federal priority and initiative and even a global phenomenon. This chapter reviews people, events, and policies that resulted in the creation of nanotechnology as a new field of science and technology. It is largely based on the author's personal experience while serving as Assistant Director for Technology R&D in the White House Office of Science and Technology Policy (OSTP) from 2003 to 2008.

The signing of the Twenty-First Century Nanotechnology Research and Development Act (the Act) on December 3, 2003, made statutory the NNI. While enactment of the legislation was a major milestone, it was the culmination of key events and the efforts of many individuals in the federal government and the scientific community (National Nanotechnology Initiative 2019b). Starting in the 1990s, federal agencies already were investing, albeit in an uncoordinated manner, in research at the nanoscale. An ad hoc interagency working group was formed, co-chaired by Mihail (Mike) Roco at the National Science Foundation (NSF) and Tom Kalil in the White House National Economic Council, and promoted by key individuals in other agencies, such as James Murday at the US Navy, who served as Executive Secretary. The group held a workshop in 1999 and with input from the broader scientific community published a report (Roco et al. 1999) outlining research opportunities and needs related to advancing nanotechnology. Based on the groundwork laid by the working group, the White House launched the NNI in 2000. The initiative was outlined by President Clinton shortly before leaving office in a speech at the California Institute of Technology that echoed a talk given by Richard Feynman in 1959, also at Cal Tech, entitled "There's Plenty of Room at the Bottom" (Feynman 1959). Feynman envisioned the ability to control matter at the atomic scale and challenged scientists to overcome the barriers. It was several decades, however, before science and engineering advances enabled Feynman's vision to be realized.

Following the launch of the new initiative at the end of the second term of the Clinton administration, the incoming George W. Bush administration embraced the new program. The interagency working group became the Nanoscale Science, Engineering and Technology (NSET) subcommittee of the National Science and Technology Council (NSTC). Starting in 2003, the NNI and nanotechnology were one of a handful of R&D priorities specified in the annual budget guidance memo from OSTP and the Office of Management and Budget (OMB) to departments and agencies (Office of Management and Budget 2003).

When considering how to organize and manage the NNI, those drafting the legislation looked to another successful multiagency federal program, the Networking and Information Technology Research and Development program (NITRD). Several processes and structures used by NITRD were adapted and incorporated in the Act, including a coordination office independent from any participating agency and reporting to OSTP, annual budget reports that include research plans, and a presidentially appointed advisory panel to review and provide expert advice. To ensure more independent assessment of the federal investment in what was an emerging and rapidly advancing field, requirement for a periodic review by the National Academies was also included in the Act.

Interagency Coordination

The NNI has benefited since its inception from strong interagency coordination made possible by a robust interagency body and a coordination office with a clear purpose. The NSET subcommittee has served as the interagency body for such coordination. The subcommittee is co-chaired by OSTP and an agency representative. The agency co-chair rotates among the participating agencies, thereby avoiding any one agency appearing to have greater influence and ensuring support among all participating agencies. The subcommittee formed subgroups or working groups to focus on certain areas deemed essential to achieving the goals of the initiative. Working groups engage individuals with relevant experience and responsibilities at participating agencies and may be established and disbanded as appropriate. In addition, coordinators have been named in cross-cutting areas "to track developments, lead in organizing activities, report periodically to the NSET subcommittee, and serve as central points of contact for NNI information in the corresponding areas" (National Nanotechnology Initiative 2019c). Coordinators also strengthen interagency engagement and interaction in a particular area, such as standards development.

After President Clinton announced the NNI, a coordination office was established, modeled after a similar office that coordinated NITRD. The National Nanotechnology Coordination Office (NNCO) was initially focused on coordinating among the multiple participating agencies; however the Act passed in 2003 enumerated the following additional roles for the office:

1. Provide technical and administrative support to the NSET subcommittee and the advisory panel
2. Serve as the point of contact on federal nanotechnology activities for government organizations, academia, industry, professional societies, state

nanotechnology programs, interested citizen groups, and others to exchange technical and programmatic information
3. Conduct public outreach, including dissemination of findings and recommendations of the advisory panel
4. Promote access to and early application of the technologies, innovations, and expertise derived from activities to agency missions and systems across the Federal Government, and to the US industry, including startup companies

Rather than funding the NNCO through appropriations within a single agency, participating agencies with budgets for nanotechnology R&D agreed to jointly fund the NNCO. The amount each agency contributed to the NNCO budget was determined by its fraction of the total nanotechnology R&D investment. This approach meant that agencies with larger budgets contributed more. It also meant that multiple agencies had "skin in the game" and an interest in seeing value from the office.

The multiagency approach to funding the NNCO also had drawbacks. It required the NNCO to deal with multiple agency contracting offices, a time-consuming effort for both sides. In addition, assessing how much each agency owed posed challenges. The total nanotechnology R&D investment is the sum of a crosscut reported to OMB of each agency's proposed R&D budget for the upcoming fiscal year. Nanotechnology is foundational and crosscutting and typically is not segregated in a single program or budget within an agency. For grantmaking agencies like NSF, a method for determining which grants are for nanotechnology R&D was needed. At a government lab, such as NIST or one of the DOD research labs, it can be difficult to categorize a research project as nanotechnology R&D. Moreover, mission agencies, such as the Department of Defense, invest in research to address mission needs, which may or may not be based on nanotechnology, making it difficult to plan future spending levels. Each agency developed a method to track nanotechnology R&D, which had to be applied consistently in order for the year-over-year estimates to be meaningful. The formula for determining how much each agency paid to support the NNCO, i.e., based on its fraction of the total investment, could incentivize agencies to be conservative when deciding whether to include projects that are only partly focused on nanotechnology R&D or otherwise difficult to categorize. On balance, however, the participating agencies have concluded that the benefits of joint support outweigh the costs.

The 2003 Act called for the NSET to prepare and update every 3 years a strategic plan for the initiative. The first strategic plan was issued in 2004 (National Nanotechnology Initiative 2004) and was organized around the following four goals:

- Maintain a world-class R&D program.
- Facilitate transfer of new technologies into products for commercial and public benefit.
- Develop educational resources, a skilled workforce, and the supporting infrastructure and tools to advance nanotechnology.
- Support responsible development of nanotechnology.

These overarching goals have remained the guideposts for the NNI. In addition, the strategic plan has identified an evolving set of objectives and actions aimed at each goal. Annual reports state progress toward the objectives through individual and joint agency activities.

In addition to a sound organizational framework and good management processes, a critical element of the NNI's success has been the individuals who serve on the NSET and as the NNCO director. NSET representatives typically are volunteers who take on the duties in addition to their full-time job. The commitment of these individuals to maximizing the advancement and utilization of nanotechnology to achieve their agency mission is evident based on the many years that most representatives serve; some current representatives have been involved since the beginning.

The NNCO director is a key position in the NNI organization. The director must serve the NSET and make sure that the work of the subcommittee is accomplished and be the liaison to OSTP on behalf of all the participating agencies. In addition, the director is the face of the NNI to the broader community. The first full-time NNCO director, appointed in early 2003, was Clayton Teague, a distinguished scientist on assignment from NIST. Teague's background in precision engineering and measurement at the nanoscale and in standards development, as well as his ability to communicate to broad audiences on all aspects of nanotechnology, made him ideally suited to the job, especially at the outset when many had questions about the potential for both good and possible harm of the new technology.

Addressing Risks

It was recognized from the beginning that nanotechnology, like all new technologies, poses certain risks. These fall into two main categories—potential to harm humans or the environment and the potential for nefarious or unethical use. Environment, health, and safety (EHS) concerns stemmed from the novel and sometimes unpredictable properties of nanomaterials and their small size, which made them hard to see and called into question the ability to filter them and to protect workers, consumers, and the environment (National Nanotechnology Initiative 2006). The possible biological applications and concerns about access to benefits in general raised ethical, legal, and societal implications (ELSI). ELSI also was a concern in the Human Genome Project and a percentage of funds appropriated for the project were set aside to invest in ELSI research starting in 1990 (National Human Genome Research Institute 2019). Although a prescribed level of funding was never specified by Congress, the Act explicitly called for the NNI to address these areas within the broad portfolio.

It was unusual when the NNI started for a research initiative to consider potential EHS risks and seek to address those risks in parallel with the basic research to advance fundamental scientific knowledge. Such a proactive approach was considered essential given the myriad applications envisioned, from medicine to food safety to environmental sensing and remediation. Addressing EHS concerns required participation of agencies with relevant expertise and responsibilities, including the Environmental Protection Agency (EPA), Food and Drug Administration (FDA), National Institute for Occupational Safety and Health (NIOSH), National Institute for Environmental Health Sciences (NIEHS), Occupational Safety and Health Administration (OSHA), and Consumer Product Safety Commission (CPSC).

The NSET subcommittee established the Nanotechnology Environmental and Health Implications working group (NEHI) in 2003 to focus on identifying potential EHS risks and the research needed to be able to assess and manage those risks. NEHI has helped to accelerate understanding related to all aspects of EHS, from monitoring exposure and dose to assessing potential toxicity compared to existing chemicals. The transparent and proactive approach was key to ensuring that agencies responsible for protecting people and the environment had accurate data. It also communicated to the broader community the measures being taken to be responsible in the development of the new technologies.

Supporting Commercialization

The name of the initiative intentionally emphasized "technology" with an eye toward the practical applications envisioned. A number of actions taken under the NNI helped to remove barriers and accelerate progress toward achieving economic and public benefits. These included the establishment of intellectual property structure and expertise in the US Patent and Trademark Office (USPTO) and collaboration with industry sectors with a stake in the development of nanotechnology.

The ability to protect intellectual property is essential to the private sector when making investments needed to commercialize a new technology. As a member of the NSET subcommittee, the USPTO kept abreast of research and was able to prepare for the rapid increase in patent applications following the launch of the NNI (Fig. 9.1). To facilitate the examination of these applications, in 2004, the USPTO created a new category for nanotechnology-related patents (Class 977) in its system for grouping patents. Working with counterparts worldwide, international patent classification systems were also modified.

To guide and accelerate nanotechnology R&D toward practical applications, NNI leaders reached out to industry sectors with an interest in nanotechnology, starting with the semiconductor and chemical industries. Convening and consulting among NNI agencies and sector-specific groups allowed the agencies to learn about basic research needs to enable certain applications and industry gained insight on current government investments. Industry interaction led to various positive outcomes.

NNI engagement with the semiconductor industry focused on the technological needs "beyond Moore's law," when scaling the current silicon-based technology to smaller sizes is not possible. The in-depth discussions and exchange of information

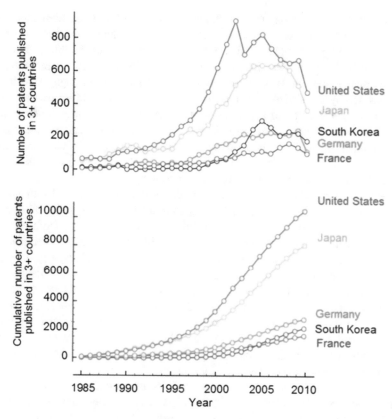

Fig. 9.1 Number of patents published in three or more countries by inventors in the United States and other leading countries in nanotechnology R&D. Source: President's Council of Advisors on Science and Technology (2014)

eventually led to the Nanoelectronics Research Initiative (NRI), a consortium of semiconductor companies that partnered with NSF and NIST to fund university research. The goal of NRI was to create technology options that could extend the performance trends associated with Moore's law into the future. NRI was a subsidiary of the Semiconductor Research Corporation (SRC) and used the SRC approach to define research needs, solicit ideas from university researchers, and connect industry experts with faculty and student researchers to guide projects and transfer results back to the companies.

Vision2020, an organization with broad participation from across the chemical industry, focused on the industry's long-term technology needs and opportunities, and partnered with NNI agencies to develop a R&D roadmap for "nanomaterials by design" (Chemical Industry Vision2020 Technology Partnership 2003). The report published in 2003 outlined research needs for advancing the ability to create nanomaterials with specific properties, including the need to understand EHS implications. It highlighted the need for R&D in manufacturing, tools for synthesis and characterization, standards, and workforce education and training. Elements of the roadmap appeared in NNI programs, including the DOE nanoscale science research centers and NSF programs related to nanomaterials and biological interactions.

A third industry partnership was with the forest product sector, represented by the American Forest and Paper Association. In 2005, the industry engaged with NNI agencies, led by the USDA Forest Service, to publish vision and technology roadmap (American Forest and Paper Association 2005). The document outlines how nanotechnology could impact the industry, e.g., to reduce the energy intensity of paper production, and how the industry could lead to advances in nanotechnology, particularly in the development of new nanomaterials based on lignocellulosic biopolymers, an abundant nanostructured plant-based material.

The hype surrounding nanotechnology in the early 2000s generated both positive and negative publicity. The eventual ability to routinely control matter of all types of matter on the scale of atoms and molecules promised everything from cures for disease to clean water and abundant, cheap renewable energy. On the other hand, the ability also could be used against the United States and its allies, e.g., to create potent chemical and biological warfare agents or hard-to-detect surveillance technologies (Clunan and Rodine-Hardy 2014). One far-off but existential scenario envisioned nanomanufacturing machines that could replicate themselves, thereby creating an exponentially growing "grey goo" (Drexler 1986). Nearer term threats of widespread use and exposure before possible harmful effects were understood seemed more likely, leading to calls for greater investment in risk research and for a precautionary approach (Balbus et al. 2005).

As for all new products that are regulated, including chemicals and materials, drugs, food, and, once available for sale, cosmetics and other consumer products, the responsible agencies act within their authorities based on data and evidence. In general, manufacturers do not want their products to cause harm to workers, customers, the public, or the environment. The potentially novel behavior of nanomaterials, however, meant that data for the same material in bulk may not apply and assumptions used when estimating risks of new chemicals may not be appropriate. Moreover, unlike many large companies, small businesses generally do not have broad in-house industrial hygiene capacity and therefore may not be aware of the latest scientific research. NEHI and the regulatory agencies engaged industry, academic experts, and nongovernmental organizations to inform decisions and raise awareness. The NNI disseminated guidelines (e.g., National Institute for Occupational Health and Safety 2009) and other documents related to EHS (National Nanotechnology Initiative 2019d).

In parallel to the development of appropriate regulations is the need to establish standards, which promote competition in the marketplace and safety of consumers and the environment. Nanotechnology is inherently difficult to define and constrain. The relative importance of quantum versus classical behavior changes gradually with size and is material specific. When does "nanomaterials" include naturally occurring nanosized particles, such as found in soot? In order for researchers, businesses, and government regulators to communicate clearly, terminology needed to be agreed upon. To ensure consistency, characterization methods needed to be established.

In the United States, the American National Standards Institute (ANSI) promotes the development of voluntary, consensus standards through transparent processes that are open to all stakeholders. ANSI also represents US interests at many international standards organizations. At the recommendation of OSTP, ANSI established the Nanotechnology Standards Panel in 2004. In 2005, the International Standards Organization (ISO) established a technical committee on Nanotechnologies (TC 229), with working groups for terminology and nomenclature; measurement and characterization; and health, safety, and environment. ANSI administers the US Technical Advisory Group (TAG) for TC 229. The first TAG Chair was Clayton Teague, who had experience in standards development and was NNCO Director at the time.

ANSI does not create standards itself; it accredits procedures of standard development organizations (SDOs). ASTM and IEEE were the first ANSI-accredited SDOs to establish technical bodies for development of nanotechnology-related standards. ASTM standards include calculation of nanoparticle sizes and size distributions and a test method for analysis of hemolytic properties of nanoparticles (ASTM 2019). ASTM also has published guides for workforce education in various areas of nanotechnology, including health and safety, materials synthesis and processing, and material properties and the effects of size. IEEE standards include test methods for characterization of electronic properties of carbon nanotubes (IEEE 2006) and a standard that is in development for large-scale manufacturing for nanoelectronics (IEEE 2015).

International Collaboration

Although the NNI was the first and largest coordinated national effort, other nations followed with similar programs, spurred in part by competition and the desire to be among the leaders of the anticipated "nanotechnology industry" (Fig. 9.2) and international scientific meetings where research results were presented proliferated. At the same time, the need for cooperation in certain areas was recognized. As mentioned earlier, international cooperation was initiated relatively quickly in the areas of patent classification and standards development.

As those responsible for EHS in counties around the world grappled with the new nanomaterials, US policy makers, led by OSTP and EPA, supported the Organization for Economic Cooperation and Development (OECD) Chemicals Committee as an ideal forum for international coordination in the area. The Chemicals Committee has worked for over four decades to protect human health and the environment while avoiding duplication of efforts, e.g., through sharing of high-quality EHS data.

In 2005, OSTP and the US State Department proposed to address topics related to nanotechnology innovation and commercialization at a government-to-government level through a newly created Nanotechnology Working Party under the OECD Committee on Science and Technology Policy (CSTP). Now part of the Working

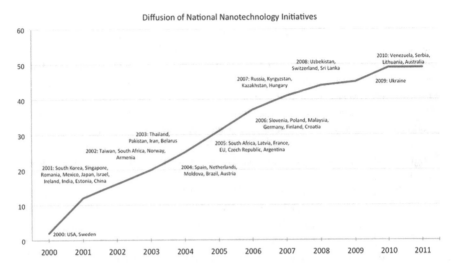

Fig. 9.2 Number of national nanotechnology initiatives established between 2000 and 2011. Source: Clunan and Rodine-Hardy (2014)

Party for Bio-, Nano- and Converging Technologies in the Directorate for Science, Technology and Innovation, the group continues to track key nanotechnology economic indicators, including number of nanotechnology firms and public sector nanotechnology R&D expenditures (OECD 2019).

Oversight and Review

The 2003 Act included provisions for periodic review of the NNI by a presidentially appointed advisory panel and by the National Academies. In July 2004, President Bush signed an executive order designating the President's Council of Advisors on Science and Technology (PCAST) as the National Nanotechnology Advisory Panel (NNAP) created by the Act. According to the Act, the NNAP is to review the NNI every 2 years. Comprising senior leaders and executives from industry and academia, PCAST was able to provide valuable input on managing and directing such a large multiagency initiative. To augment PCAST expertise in key technical areas relevant to nanotechnology, a technical advisory group (TAG) of some 50 experts was created. The "nano TAG" was surveyed and consulted by PCAST to inform its assessment and recommendations. The Obama and Trump administrations also designated PCAST as the NNAP. In addition to the advisory panel review, the Act calls for the National Academies to review the NNI, including 13 specific aspects of the program funding, organization, strategy, and management, every 3 years. The first National Academies review was published in 2006 (National Research Council 2006).

Outside expert assessment and input is a valuable mechanism for keeping a program like the NNI on track. However, by being so specific in what each assessment must address, the advisory panel and National Academies committees are effectively constrained with insufficient time or resources to consider other topics that might be useful for managing and achieving the goals of the program. When the Act was passed in 2003, nanotechnology was new and very dynamic and relatively frequent detailed assessments were appropriate. As the NNI matures, the type of review that can help keep the program on the leading edge will evolve. Legislation enacted in 2017 amended the schedule for NNAP and National Academies reviews to every 4 years and extended the interval to 5 years for updating the strategic plan. However, the nature of the assessment has not been revised or made more flexible.

Concluding Thoughts

The NNI has been a poster child for how a government-led program can accelerate progress in a new area of technology by a coordinated investment in advancing knowledge combined with practical approaches to addressing potential risks and barriers to commercialization. Today, the NNI involves 20 federal agencies that collectively invest approximately $1.5 billion annually. The program's success stemmed from a combination of factors.

- Visionary founders: The NNI benefited from significant contributions by many individuals, and in particular by Mike Roco at NSF, who had vision and was indefatigable in his efforts to create and sustain the initiative. In addition, with his background in precision measurement and experience in related standards development Clayton Teague was the right person at the right time to be the first full-time NNCO director. Support from OSTP and OMB ensured that agencies prioritized nanotechnology in the critical early years.
- Good timing: Technical advances in the ability to characterize and manipulate matter at the scale of atoms and molecules set the stage for individual and coordinated agency investments that could address agency missions.
- Robust coordination: Establishing the NNCO provided essential support for coordination among agencies and between the federal government and other entities, including the private sector and the public.
- Attention to risk: The NNI included research and other activities aimed at understanding EHS and other societal implications of nanotechnology. To accomplish objectives in this area, regulatory agencies were engaged early in the program.

It is difficult to answer the question, "What would have happened if there hadn't been an NNI?" Nanoscale science would have advanced, though perhaps not at the same rate. It can be argued that nanotechnology served as the linchpin for converging multidisciplinary research, along with biotechnology and information technology. The emphasis on multidisciplinary research forced teams of researchers to collaborate. Supporting multidisciplinary programs required funding agencies to allow crossing of silos and boundaries among discipline-based programs and offices.

The NNI was unusual for the breadth of engagement and participation. Among the 20 federal agencies that joined the program are several that do not have significant research budgets but play essential roles in promoting innovation and managing risks. Enduring relationships among NSET representatives led to multiagency programs that would not have happened otherwise. Strategic engagement with industry at an early stage was invaluable in guiding researchers and research programs. To ensure that the United States had the necessary talent, the NNI supported nanotechnology education at all levels—both formal and informal.

What lies ahead for nanotechnology? Some nations have ended their separate nanotechnology programs. In part due to the 2003 Act, the US NNI continues to exist and supports activities aimed at growing nano-enabled technology—through research and education, and by focusing the program in areas where it can have a substantive impact. The NNI has built a foundation of knowledge and capability that can advance new priority areas, such as quantum information science.

There is still "plenty of room at the bottom." The design and creation of complex nanostructured materials and systems is in its infancy. The ability to manipulate atoms precisely remains largely a painstaking process and scaling to industrial manufacturing levels is a challenge. Based on the achievements to date, the power of the NNI to cross boundaries and build coalitions among federal agencies and with the private sector to address challenges—in energy, healthcare, nanoelectronics, aerospace, and myriad other applications—is sure to lead to even greater advances and opportunities in the next 15 years.

References

American Forest and Paper Association. (2005) *Nanotechnology for the forest products industry: Vision and technology roadmap.* Retrieved January 20, 2019 from http://www.tappi.org/content/pdf/nanotechnology_roadmap.pdf.

ASTM (2019) *Nanotechnology standards.* Retrieved January 20, 2019, from https://www.astm.org/Standards/nanotechnology-standards.html.

Balbus, J., Denison, R., Florini, K., & Walsh, S. (2005). Getting nanotechnology right the first time. *Issues in Science and Technology, 21*(4), 65–71.

Binnig, G., Quate, C. F., & Gerber, C. (1986). Atomic force microscope. *Physical Review Letters, 56*, 930–933.

Buffat, P., & Borel, J. P. (1976). Size effect on the melting temperature of gold particles. *Physical Review A, 13*, 2287–2298. Retrieved from https://doi.org/10.1103/PhysRevA.13.2287.

Chemical Industry Vison2020 Technology Partnership. (2003). *Chemical industry R&D roadmap for nanomaterials by design: From fundamental to function.* Retrieved from https://www.nanowerk.com/nanotechnology/reports/reportpdf/report17.pdf.

Clunan, A. L., & Rodine-Hardy, K. (2014). *Nanotechnology in a globalized world: Strategic assessments of an emerging technology.* U.S. Naval Postgraduate School, Project on Advanced Systems and Concepts for Countering WMD.

Drexler, K. E. (1986). *Engines of creation.* Garden City, NY: Anchor Press/Doubleday.

Feynman, R. (1959, December 29). *There's plenty of room at the bottom*. Paper presented at the annual meeting of the American Physical Society, California Institute of Technology, Pasadena.

Gao, F., & Gu, Z. (2016). Melting temperature of metallic nanoparticles. In M. Aliofkhazraei (Ed.), *Handbook of nanoparticles*. Cham: Springer. https://doi.org/10.1007/978-3-319-15338-4_6.

IBM. (2009). *IBM celebrates 20th anniversary of moving atoms*. Retrieved January 16, 2019, from https://www-03.ibm.com/press/us/en/pressrelease/28488.wss.

IEEE. (2006). *Standard test methods for measuring electrical properties of carbon nanotubes*. IEEE Standard 1650-2005. Retrieved January 20, 2019, from https://standards.ieee.org/standard/1650-2005.html.

IEEE. (2015) *Large scale manufacturing for nanoelectronics*. IEC/IEEE International Standard 62659-2015. Retrieved January 20, 2019, from https://standards.ieee.org/standard/62659-2015.html.

National Human Genome Research Institute. (2019). *The ethical, legal, and societal implications research program*. Retrieved January 20, 2019, from https://www.genome.gov/10001618/the-elsi-research-program/#al-1

National Institute for Occupational Health and Safety. (2009). *Approaches to safe nanotechnology* (pp. 2009–2125). DHHS (NIOSH) Publication.

National Nanotechnology Initiative. (2004). *The national nanotechnology initiative strategic plan*. Retrieved January 20, 2019, from https://www.nano.gov/sites/default/files/pub_resource/nni_strategic_plan_2004.pdf

National Nanotechnology Initiative. (2006). *Environmental, health, and safety research needs for engineering nanomaterials*. Retrieved January 20, 2019, from https://www.nano.gov/sites/default/files/pub_resource/nni_ehs_research_needs.pdf

National Nanotechnology Initiative. (2019a). *What it is and how it works*. Retrieved January 20, 2019, from https://www.nano.gov/nanotech-101/what.

National Nanotechnology Initiative. (2019b). *Nanotechnology timeline*. Retrieved January 20, 2019, from https://www.nano.gov/timeline.

National Nanotechnology Initiative. (2019c). *Working groups and coordinators*. Retrieved January 20, 2019, from https://www.nano.gov/about-nni/working-groups.

National Nanotechnology Initiative. (2019d). *Environmental, health, and safety-related documents*. Retrieved January 20, 2019, from https://www.nano.gov/node/1164.

National Research Council. (2006). *A matter of size: Triennial review of the national nanotechnology initiative*. Washington, DC: The National Academies Press. Retrieved January 20, 2019, from https://doi.org/10.17226/11752.

OECD. (2019). *Key nanotechnology indicators*. Retrieved January 20, 2019, from http://www.oecd.org/sti/emerging-tech/nanotechnology-indicators.htm.

Office of Management and Budget. (2003). *FY 2005 interagency research and development priorities*. Retrieved January 20, 2019, from https://www.whitehouse.gov/sites/whitehouse.gov/files/omb/memoranda/2003/m03-15.pdf.

President's Council of Advisors on Science and Technology. (2014). *Report to the president and congress on the fifth assessment of the national nanotechnology*. Retrieved January 20, 2019, from https://www.nano.gov/sites/default/files/pub_resource/pcast_fifth_nni_review_2014.pdf.

Roco M, Williams S, Alivisatos P, Eds. (1999). *Nanotechnology research directions: IWGN workshop report*. Vision for nanotechnology R&D in the next decade. Retrieved January 20, 2019, from https://www.nano.gov/sites/default/files/pub_resource/research_directions_1999.pdf.

Royal Swedish Academy of Sciences. (1986). Retrieved January 20, 2019, from https://www.nobelprize.org/prizes/physics/1986/press-release/.

U.S. Congress. (2003). *Public law 108-153: 21st century nanotechnology research and development act.*. Retrieved January 20, 2019, from https://www.congress.gov/108/plaws/publ153/PLAW-108publ153.pdf.

Index